The Revolution in Physics

LOUIS DE BROGLIE

THE REVOLUTION IN PHYSICS

A Non-mathematical Survey of Quanta

Translated by Ralph W. Niemeyer

GREENWOOD PRESS, PUBLISHERS
NEW YORK

Reprinted by arrangement
with Farrar, Straus, & Giroux, Inc.

First Greenwood Reprinting 1969

SBN 8371-2582-0

PRINTED IN UNITED STATES OF AMERICA

Contents

The Revolution in Physics

Introduction | The Importance of Quanta

1. Why It Is Necessary to Know About Quanta

AMONG THOSE who will glance at the cover of this little book, many no doubt will be frightened away by the sight of that mysterious word: quanta. The general public does have some vague ideas—yes, often very vague—about the theory of relativity concerning which there has been a great deal of talk in the last several years; but the said general public has, I believe, few ideas—even vague ones—about the quantum theory. It must be said, this is excusable, for quanta are a very mysterious thing. In my case, I was about twenty when I began to work with them and I have now been pondering over them a quarter of a century; very well, I must humbly confess that if in the course of these meditations I have come to understand some of their aspects a little better, I do not yet know exactly what is hidden behind the mask which covers their true face. Nevertheless, it seems to me one thing can be asserted: despite the importance and the extent of the progress accomplished by physics in the last centuries, as long as the physicists were unaware of the existence of quanta, they were unable to comprehend anything of the profound nature of physical phenomena for, without quanta, there would be neither light nor matter and, if one may paraphrase the Gospels, it can be said that without them "was not anything made that was made."

We can then imagine the essential shift in direction which the course of the development of our human science underwent the day when quanta, surreptitiously, were introduced into it. On that day, the vast and grandiose edifice of classical physics found itself shaken to its very foundations, without anyone very clearly realizing it at first. In the history of the intellectual world there have been few upheavals comparable to this.

Only today are we beginning to be able to measure the extent of the revolution which has come to pass. Faithful to the Cartesian ideal, classical physics showed us the universe as being analogous to an immense mechanism which was capable of being described with complete precision by the localization of its parts in space and by their change in the course of time, a mechanism whose evolution could in principle be forecast with rigorous exactness when one possessed a certain amount of data on its initial state. But such a conception rested on several implicit hypotheses which were admitted almost without our being aware of them. One of these hypotheses was that the framework of space and time in which we seek almost instinctively to localize all of our sensations is a perfectly rigid and fixed framework where each physical event can, in principle, be rigorously localized independently of all the dynamic processes which are going on around it. Thereupon, all the evolutions of the physical world are necessarily represented by modifications of local states of space in the course of time, and that is why in classical science dynamic quantities such as energy and momentum appeared as derived quantities constructed with the aid of the concept of velocity, kinematics thus serving as the basis for dynamics. From the point of view of quantum physics, it is entirely otherwise. The existence of the quantum of action, to which we must so often return in the course of this work, does imply a kind of incompatibility between the point of view of a localization in space and time and the point of view of a dynamic evolution; each of these points of view is capable of being used in a description of the actual world, but it is not possible to adopt both simultaneously in all their rigor. Exact localization in time and space is a sort of static idealization which excludes all evolution and all dynamism; the idea of a state of motion taken in all its purity is on the contrary a dynamic idealization contradictory in principle with the concepts of position and instant. A description of the physical world in the quantum theory can only be made by using to a greater or lesser degree one or the other of these two contradictory images: there results then a sort of compromise and the famous Uncertainty Relations of Heisenberg tell us in what measure this compromise is possible. Among other consequences, it follows from the new ideas that kinematics is no longer a science having a physical meaning. In classical mechanics it was permissible to study displacements in space by themselves and to define thereby velocities and accelerations without bothering about the manner in which these displacements are materially realized: from this abstract study of motion, one then progressed to dynamics by introducing several new physical principles. In quantum mechanics a similar division of the subject is no longer admissible in principle since the spatio-temporal localization which is the basis of kinematics is acceptable only to a degree which depends on the dynamic conditions of motion. We shall see later why it is nevertheless perfectly legitimate to make use of kinematics when studying large scale phenomena; but for phenomena on the atomic scale where quanta play a predominant role, it can be said that kinematics, defined as the study of motion made independently of all dynamic considerations, completely loses its significance.

Another implicit hypothesis underlying classical physics is the possibility of making negligible by appropriate precautions the perturbations which are exerted on the course of natural phenomena by the scientist who, in order to study them with precision, observes them and measures them. In other words, it is assumed that in well-conducted experiments the perturbation in question can be made as small as one wishes. This hypothesis is always sensibly realized for large-scale phenomena, but it ceases to be so in the atomic world. Indeed, as the fine and profound analyses of Heisenberg and Bohr have shown, it follows from the existence of the quantum of action that all attempts to measure a characteristic quantity of a given system has the effect of changing in an unknown fashion the other quantities associated with this system. More precisely, any measurement of a quantity which permits the exact localization of a system in space and time has the effect of changing in an unknown fashion a quantity which is conjugate with the first and which serves to specify the dynamic state of the system. In particular, it is impossible to measure at the same time and with precision any two conjugate quantities. We can then understand in what sense it can be said that the existence of the quantum of action makes a spatio-temporal localization of the parts of a system incompatible with attributing a well-defined dynamic state to that system, since, in order to localize the parts of a system, it is necessary to know exactly a series of quantities, the knowledge of which excludes that of the conjugate quantities relative to the dynamic state, and conversely. The existence of quanta imposes a lower limit of a very definite kind on the perturbations which a physicist can exert on the systems he is studying. Thus, one of the hypotheses which served implicitly as a basis of classical physics is contradicted and the consequences of this fact are considerable.

From the above it follows that we are never able to know

with precision more than half the quantities, the knowledge of which would be necessary for an exact description of a system in accordance with classical ideas. The value of a quantity characterizing a system is indeed that much more uncertain when that of its conjugate quantity is more exactly known. From this springs an important difference between the old and the new physics in so far as the determinism of natural phenomena is concerned. In the old physics, the simultaneous knowledge of the quantities fixing the position of the parts of a system and of the conjugate dynamic quantities permitted, at least in principle, a rigorous calculation of the state of the system at a later instant. Knowing with precision the values x_0, y_0 . . . of the quantities characterizing a system at an instant t_0, we could predict without ambiguity what values x, y . . . would be found for these quantities if they were determined at a later instant t. This resulted from the form of the basic equations of the physical and mechanical theories and from the mathematical properties of these equations. This possibility of a rigorous forecasting of future phenomena starting from present phenomena, a possibility implying that the future is in some manner contained in the present and adds nothing to it, constituted what has been called the determinism of natural phenomena. But this possibility of a rigorous forecasting requires the exact knowledge at the same instant of time of the variables of spatial localization and of the conjugate dynamic variables; now, it is precisely this knowledge which quantum physics considers impossible. Hence there has resulted a considerable change in the manner in which physicists (or at least a great many of them) today conceive the power of forecasting in physical theories and of the concatenation of natural phenomena. Having determined the values of the quantities characterizing a system at an instant t_0 with the uncertainties with which they are necessarily affected in quantum theory, the physicist can not predict exactly what the value of these quantities will be at a later instant; he can only state what the probability is that a determination of these quantities at a later instant *t* will furnish certain values. The bond between successive results of measurements, which convey for the physicist the quantitative aspect of phenomena, is no longer a causal bond consistent with the classical deterministic scheme, but rather a bond of probability which alone is compatible with the uncertainties that, as we have explained above, derive from the very existence of the quantum of action. And here is an essential modification of our conception of physical laws, a modification of which all the philosophical consequences are still far, we believe, from being fully realized.

Two ideas of considerable general application have come out of the recent evolution of theoretical physics: that of complementarity in the sense of Bohr and that of the limitation of concepts. Bohr was the first to observe that in the new quantum physics, in the form which the development of wave mechanics has imparted to it, the ideas of corpuscles and waves, of spatio-temporal localization and well-defined dynamic states are "complementary"; by that he means that a complete description of observable phenomena requires that these concepts be employed in turn, but that these concepts nevertheless are in a sense irreconcilable, the images that they furnish never being simultaneously applicable *in toto* to a description of reality. For example, a great number of observed facts in atomic physics can be treated simply by invoking only the idea of corpuscles so that the use of this idea can be considered indispensable to the physicist; likewise, the idea of waves is equally indispensable for the description of a great number of phenomena. If one of these two ideas were rigorously applied to reality, it would completely exclude the other. But it is found that both of them in fact are useful in a certain degree for the description of phenomena and that, despite their contradictory character, they should be used alternately depending on the situation. It is the same for the ideas of a spatio-temporal localization and of a well-defined dynamic state: they too are "complementary" as are the ideas of corpuscles and waves to which they are moreover closely allied as we shall later see. It can be asked how it is these two contradictory images never happen to collide head-on. We have already indicated the reason for this: the two complementary images can not collide head-on because it is impossible to determine simultaneously all the details which would permit making these two images entirely precise and this impossibility, which is expressed in analytical language by the Uncertainty Relations of Heisenberg, rests squarely on the existence of the quantum of action. In this way the important role played by the discovery of quanta in the evolution of contemporary theoretical physics appears in all its clarity.

To complementarity in the sense of Bohr, the limitation of concepts is closely allied. Simple images, such as that of a corpuscle, of a wave, of a point in space, of a well-defined state of motion, are, in short, abstractions, idealizations. In a great number of cases, these idealizations are found to be approximately realized in nature, but they nevertheless have a limited application; the validity of each of these idealizations is limited by the validity of the "complementary" idealization. Thus corpuscles can be said to exist since a great number of phenomena can be interpreted by invoking their existence. Nevertheless, in other phenomena the corpuscular aspect is more or less hidden and it is an undulatory aspect that is revealed. These more or less schematic idealizations which our mind constructs are capable of representing certain aspects of things, but they have their limits and can not incorporate into their rigid forms all the richness of reality.

We do not wish to continue too long this first survey of the new viewpoints, which has let us get a glimpse of the development of quantum physics. We shall have the occasion in the course of this volume to take up again all these questions one by one, completing them and investigating them thoroughly. What we have said will suffice to show the reader how deep the interest of the quantum theory is: it has not only stimulated atomic physics, which is the most really alive and zealous branch of the physical sciences, but it has also incontestably broadened our horizons and introduced a number of new ways of thinking, deep traces of which will without doubt remain in the future expansion of human thought. For this reason, quantum physics should interest not only specialists: it merits the attention of all cultivated men.

2. *Classical Mechanics and Physics Are Approximations*

We now should like to examine quickly what value the whole of classical mechanics and physics retains in the eyes of a quantum physicist. Of course, they retain practically all their value in the domain of facts for which they were created and in which they have been verified. The discovery of quanta does not prevent the laws of the fall of heavy bodies or those of geometrical optics from remaining valid. Each time a law has been verified in an incontestable manner to a certain degree of approximation (all verification carries with it a certain degree of approximation), we have a definitely acquired result which no later speculation is able to undo. If it were not thus, no science would be possible. But it can very well be that, in the light of new experimental facts or of new theoretical conceptions, we are led to consider previously verified laws as being only approximate, that is to assume that, if the precision of the verification were indefinitely increased, the laws would not be more exactly verified. This has happened many times in the course of the history of science. Thus the laws of geometrical optics—for example, the rectilinear propagation of light—although having been verified with precision and at first regarded as rigorously true, were seen to be only approximations that day when the phenomena of diffraction and the wave character of light were discovered. It is just by this process of successive approximations that science is capable of progressing without contradicting itself. The structures that it has solidly built are not overthrown by subsequent progress, but rather incorporated into a broader structure.

It is in this way that classical mechanics and physics can be regarded as entering into the framework of quantum physics. Classical mechanics and physics were built up in order to give an account of phenomena which operate on an ordinary, human scale and they are also valid for larger scales, the astronomical ones. But, if one descends to the atomic scale, the existence of quanta is going to limit their validity. Why should this be so? Because the value of the quantum of action, measured by Planck's famous constant, is extraordinarily small in relation to our usual units, i.e., in relation to the quantities that are found on our scale. The perturbations introduced by the existence of quanta, in particular the uncertainties of Heisenberg, are a great deal too small in the usual conditions on our scale to be perceptible; they are indeed a great deal smaller than the inevitable experimental errors which affect the verification of the classical laws.

In the light of the quantum theories, classical mechanics and physics seem then not to be rigorously exact in principle, but their inexactness is completely masked in usual cases by experimental errors in such a way that they constitute excellent approximations for phenomena on our scale. Thus we meet again the customary process of scientific progress: well-established principles, well-verified laws are conserved, but they can be considered valid only as approximations for certain categories of facts.

Perhaps, in the presence of this validity of classical mechanics or physics for facts on our ordinary scale where quanta do not intervene, one might be tempted to say: "In short, quanta do not have all the importance you attribute to them since, in all the immense domain where classical mechanics and physics are valid, a domain which covers in particular that of practical applications, quanta can be completely ignored." Such a way of looking at this matter does not seem to us justified. First of all, in a domain so alive, so important, so full of future possibilities,[1] as that of atomic and nuclear physics, quanta do play an essential role and it is totally impossible to interpret the phenomena without appeal to them. Then, in macroscopic physics, quanta, although hidden by virtue of their smallness through an unavoidable lack of precision in measurement are however there, and their existence entails in principle all the consequences we have enumerated: if in practice these consequences have no appreciable influence, that does not detract in any way from their general, philosophical application. The knowledge and study of the quantum of action is one of the essential bases of natural philosophy today.

[1] This text was written ten years ago. The recent realization of the atomic bomb shows very well what deep consequences the progress of atomic and nuclear physics can have in the field of practical applications. (Note added in 1946)

Chapter 1 | Classical Mechanics

1. Kinematics and Dynamics

By no means do we have the intention of attempting, in the space of a very short chapter, an analysis or critique, even in summary, of the principles of classical mechanics. To do one or the other an entire volume would not suffice and, besides, this has already been done by eminent scholars. We only wish to emphasize several particular points, which are just those that seem interesting to us from the point of view of the subject we are treating.

Treatises on rational mechanics distinguish two studies of a very different character: the study of kinematics and that of dynamics (of which statics is just a special case). It is important to reflect a little on this division of classical mechanics for it rests on hypotheses which, as we have already briefly indicated in the introduction, no longer seem justified from a quantum point of view. What, in fact, is kinematics and why is its study put before that of dynamics? Kinematics is by definition the study of the motions within a framework of three-dimensional space which are realized in the course of time, a study made independently of the physical laws of these motions. It seems entirely natural to put the examination of kinematics before that of dynamics for it seems absolutely logical to study *in abstracto* the different kinds of motion in space before asking for what reasons and according to what laws such and such a motion is effectively produced in such and such a circumstance. But however natural this point of view may seem, it still carries with it an hypothesis which until modern times had escaped the most acute minds. Indeed it obviously is permitted a mathematician to study motions in a three-dimensional space as a function of a parameter which will be identified with time. But it is a question of knowing if this abstract study is necessarily usable, as was assumed without discussion up to now, when we want to take up the study of the actual motion of physical objects. The classical transition from kinematics to dynamics in effect implies the hypothesis that the localization of physical objects in the abstract framework of three-dimensional space and time is possible independently of the intrinsic properties of these physical objects, of their mass, for example. Now it is certain that the material bodies which surround us and which *are* on our ordinary physical scale, allow themselves to be readily located in space and time. It is the properties of these bodies, and especially of solid bodies, which have led us to imagine a space of three dimensions in which we suppose them to be distributed and it is the motions of these bodies which have permitted us to give a precise definition to the flow of time and to its measurement. Hence, it is very natural that, for these bodies, the method followed by rational mechanics should flourish and should lead to the great successes familiar to us. But it is too rash an extrapolation to suppose, as at the beginning of the development of atomic physics, that the possibility of locating physical objects in a three-dimensional space and in time can be extended without modification to the elementary particles of matter, i.e., to extraordinarily light objects. In reality, the classical conceptions of space and time are no longer valid for these ultimate objects and we can utilize them only at the price of restrictions and uncertainties which form the strangest aspect of the quantum theory. Later we shall have to examine this question more in detail: for the moment it will suffice to have noted on what implicit hypothesis, the validity of which is assured only for objects on our scale, rests the method followed by classical mechanics for the description and study of the motions of material bodies.

2. *Newton's Laws of the Dynamics of a Material Point*

Having taken as basic the possibility of representing exactly physical objects in a framework of space and time, classical mechanics starts with the simplest case, i.e., that of a physical object endowed with mass and of negligible dimensions. This schematic picture of an elementary grain of matter which rational mechanics thus puts at the beginning of the exposition of the laws of dynamics is quite in accord with the conception of a discontinuous structure of matter, and when physicists were seeking a half century ago to picture matter as an aggregate of elementary corpuscles in motion, they very naturally found in the dynamics of a material point the instrument they needed for their theoretical investigations.

The dynamics of a material point starts from the principle of inertia, according to which a material point, not subjected to any external action, keeps its same state of motion (or of rest) in the course of time. At least, this statement is exact when the motion of the material point is referred to certain systems of coordinate axes called "Galilean systems," for example, to a system bound to the system of fixed stars. The preferential role of these Galilean systems was interpreted in the following manner: since three-dimensional space in which physical objects are located is conceived as having an absolute sense, the Galilean systems of axes are those which are at rest or in uniform, rectilinear motion with respect to absolute space.

According to the principle of inertia, the motion of a free material point is a uniform, rectilinear motion of constant velocity which can moreover be a state of rest if the velocity is zero. Therefore it is very natural to suppose that if the material point is subjected to the action of a force, the effect of this force is to change the velocity. Then the simplest hypothesis is to assume that the instantaneous variation of the velocity is proportional to the force, the coefficient of proportionality being just that much smaller when the given material point opposes a change in velocity with greater inertia. We are therefore led to characterize the material point by a coefficient of inertia, its mass: the fundamental law of the dynamics of a material point will then be that the acceleration of a material point is equal at each instant to the quotient of the force acting on it by its mass. It will be noticed that the coefficient of mass, whose role is to characterize the material point from a dynamic point of view, is introduced *a posteriori*, the existence of a well-defined position, trajectory, velocity and acceleration of the material point being assumed *a priori* in conformity with the method that assumes the anteriority of kinematics over dynamics.

The equations of the classical dynamics of a material point state then that the product of the mass of the point by any of the rectangular components of its acceleration is equal to the corresponding component of the force. If the force is supposed known at each point for all values of the time, we then have to solve a system of three differential equations of second order with respect to time where the unknowns are the coordinates of the point. A well-known theorem in analysis tells us that if we know at a certain initial time the values of the coordinates and their derivatives with respect to time, the solution of the system of equations in question is entirely determinate, i.e., if the position and velocity of a material point are supposed known at a given instant, all its subsequent motion is entirely predictable. This result expresses the fact that the classical dynamics of a material point is entirely in accord with the postulate of physical determinism, a postulate according to which the future state of the material world must be entirely predictable when a certain amount of data on its present state is known.

It is interesting to make another remark here. Since the material point is supposed to be a geometric point, its trajectory is a line which explores only a one-dimensional continuum in three-dimensional space. At each point in its trajectory, the material point finds a value of the force which determines its motion during the next infinitesimal instant and hence it explores the field of force only along its trajectory. Nevertheless, it can be said that the motion depends in reality on the field of force in the region which immediately surrounds the trajectory. For in all physical problems the field of force generally varies in a continuous fashion in space in a way that the value of the force at a point in the trajectory is not independent of its values in the immediate neighborhood of the trajectory. This is clearly seen in the frequently found case where the force is derived from a potential, i.e., when the force at each point is equal to the gradient of a certain function of position: in fact, the definition of the gradient supposes that the point of application in space of the force being considered is made to vary an infinitesimal amount, and hence the force at each point of the trajectory seems truly dependent on the values of the potential in the region of space which immediately surrounds the trajectory. The principle of Least Action, of which we shall speak later on, leads to the same conclusion, for it determines the actual trajectory of a material point, the one which is really described according to the laws of dynamics by comparing it with other virtual trajectories that are infinitely near, and this implies for the determination of the motion the influence of the entire region of space infinitely near the actual trajectory. But, of course, in classical mechanics topological accidents which might exist in space at finite distances from the path of a material point can in no way influence its motion. For example, let us put a screen pierced by a hole in the path of the material point. If the path passes through the center of the hole, it will in no way be disturbed by the topological accident which the screen constitutes: if, on the contrary, the path passes infinitely near the edge of the screen, it will be disturbed and one says in popular language that the corpuscle has grazed the edge of the screen. But it is inconceivable in classical mechanics that the motion of a material point traversing the hole in question would depend on the fact of there being or not being other holes in the screen at a finite distance from the first. One can immediately understand the importance of these remarks as they concern a corpuscular interpretation of Young's interference experiment and one has a foreknowledge of all that wave mechanics must bring up again on this point.

The equations of classical mechanics of a material point have led us to consider two quantities of a dynamic nature which characterize the motion of a material point. The first of these quantities is of a vectorial nature: it is momentum which classical mechanics defines as the product of the mass of a material point by its velocity. The importance of this quantity comes from the equations of motion themselves, for these equations can be expressed by saying that the vector "derivative of momentum with respect to time" is constantly equal to the force acting on the material point. As can be seen, the classical theory derives this dynamic quantity from a kinematic quantity "velocity" with the help of a simple multiplication by the factor of mass, but nevertheless a great difference in nature is felt between the velocity and the momentum since the second of these quantities brings into play the intrinsic dynamic properties of a given material point.

The same thing applies to the second quantity to which we

made allusion above: energy. This is a scalar quantity which plays an essential role in the very important case where the force is derived from a "potential" function. If the potential does not vary with time at each point, it immediately follows from the equations of motion that a certain quantity defined at each instant by the state of the material point remains constant during the course of the motion: this quantity is equal to one half the product of the mass by the velocity squared plus the value of the potential at the place where the material point is located, i.e., it is equal to the sum of the kinetic and potential energy. Thus in a permanent field of force arising from a potential (conservative field) the total energy defined as we have just recalled remains constant: it is, in mathematical language, a first integral. Here again we find for energy an expression formed with the aid of the kinematic concept of velocity to which the specifically dynamic concepts of mass and potential are joined. (The latter is directly related to force.) It is unnecessary to recall the importance that the notion of energy, in transcending the bounds of mechanics, has acquired in all of physics. Just as the energy is constant when the derivative of the potential with time is always zero, so one of the components of the momentum remains constant when the derivative of the potential with respect to the corresponding coordinate always remains zero. This remark shows a certain relationship between energy and the components of the momentum, energy corresponding to the coordinate time while the components of the momentum correspond to the space coordinates. This relationship has been made precise by the theory of relativity which considers the energy and the three components of the momentum as forming the components of a vector in space-time: the world-force 4-vector.

The mechanics of a material point also introduces other important quantities. Among such are the components of the moment of momentum (or moment of rotation) of the material point around a fixed point, and these also are expressed by adding the dynamic idea of mass to the kinematic concepts of position and velocity. It is known that these components are first integrals when the field of force is central with respect to the given fixed point: the importance of this case in celestial mechanics is well known.

To summarize, in classical theory the essential dynamic quantities are constructed by starting with the kinematic notions of position and velocity onto which the properly dynamic notions of mass and potential are superimposed. We shall see that these things are presented in a much different fashion in present day quantum theories.

3. The Dynamics of Systems of Material Points

In the dynamics of a material point we suppose that the field of force is given at each point and at each instant. But in the concepts of classical mechanics the field of force which acts on a material point is itself created by other material points. Thus we are naturally led to consider aggregates of material points which interact on each other and to determine the possible motions of such aggregates. At first glance the problem may seem complicated for each material point of the system is displaced under the influence of the actions of the other material points of the system and this displacement has the effect of changing the forces that a given material point exerts on the others. Nevertheless, from an analytical point of view the problem is presented simply: we shall write, for each material point, that at each instant the product of its mass by its acceleration is equal to the instantaneous force which acts on it, a force which naturally depends on the positions of all the material points of the system. Thus for a system formed of N material points we obtain an array of 3N differential equations of second order with respect to time between the 3N coordinates of the N material points. Mathematical analysis then shows that the solution of this array of equations is entirely determinate if we know at a certain initial instant the position and velocity of all the points of the system. Thus the mechanistic determinism already established for the motion of a single material point is found to be extended to a system of material points.

The study of the motions of systems of material points becomes very much simplified through the consideration of the center of gravity which is, as we know, the weighted mean position of all the points of the system. This point turns out to have a simple motion, one that is rectilinear and uniform, if the system is not subjected to external actions. This results from a general property of forces introduced into mechanics, a property which is expressed by the principle of the equality of action and reaction. According to this principle, the force exerted by a material point A on a material point B is equal and opposite to the force which B exerts on A. When there exists a potential energy in the system, it amounts to assuming that this potential energy depends only on the mutual distances of the material points, a very natural hypothesis from a physical point of view. Rational mechanics in this way permits us to resolve into two parts the problem of determining the motion of a system by studying first the motion of the center of gravity, then the motion of the system around its center of gravity. A whole series of well-known theorems facilitates this study.

The momentum of a system of material points is defined very simply as being the (geometric) sum of the momenta of the constituents. Its expression is then given as the sum of the products of each mass by the corresponding velocity and this expression always uses the concept of velocity. As for the energy of the system, it is always comprised of a kinetic part which is the sum of the kinetic energies of the different material points: its expression is given by one-half the sum of the product of each mass by the square of the corresponding velocity. But if the system is conservative, the energy is also comprised of a potential energy which, too, is divided into two parts: the first is the sum of the potential energies that each material point possesses in the external field to which the whole system is subjected, if there is one; the second part of the potential energy, the only part which exists if there is no external force, is the mutual energy of the material points and it is equal to the sum of the mutual potential energies of the points taken two by two. It is most remarkable that the mutual potential energy is not resolved into the sum of the potential energies attributed individually to each material point. There is a kind of "pooling" of potential energy for each pair of points in interaction and consequently a kind of diminution of the individuality of the material points which is that much more accentuated when the interaction itself is accentuated. This pooling of a part of the potential energy is a feature which characterizes systems of material points in interaction and distinguishes them, for example, from assemblages of material points not in interaction placed in a given external field.

The dynamics of systems of material points has served as a basis for the dynamics of solid bodies, for the latter can be considered as being formed of material points whose distances are constrained to remain fixed by the mutual forces becoming extremely large as soon as the mutual distances tend to depart from their normal values. The fact that in a single solid body the mutual distances are invariable permits its position to be characterized at each instant by 6 parameters only, for example —the three coordinates of an arbitrary point of the body and the three angles which fix the orientation of the body around that point. If the problem involves several solid bodies, subject to some restraining bonds between the various bodies, it is advisable to introduce a greater number of parameters, but the equations of motion can always be written by starting from those of the material points which are supposed to constitute the solid bodies.

Thus, anticipating the progress of atomic physics, the mechanics of solid bodies was developed by assuming a discontinuous constitution of matter. Here, it will be useful to make a remark. In our ordinary experience it is large-scale bodies and not material points which are observed by us: in particular, the greater part of the operations of measuring time and space, which permit us to give precision to our study of the progress of phenomena, make use of solid bodies. Therefore it is notions drawn from observations on large-scale bodies and especially from those on solid bodies which aid us in defining the laws of motion of material points and, once these laws are admitted, we can re-deduce the mechanical properties of solid bodies by considering them as being formed of material points. Surely, there is nothing contradictory here, but it is on the whole a rash hypothesis to suppose that the notions of space and time acquired and made precise through observation of solid bodies are applicable without modification to elementary corpuscles, to material points: one could very well suppose that these notions would have need of profound modifications in order to be applied to elementary corpuscles, the only condition which is really imposed on us being that the properties of the elementary corpuscles ought to let us recover for systems of very many corpuscles the known properties of material bodies (especially of solid bodies) and the usual definitions of space and time. This point of view, whose importance has recently been stressed by Jean-Louis Destouches, does not perhaps constitute a real objection to the method followed by classical rational mechanics, for the material point could be defined there not as an elementary corpuscle, but as a little fragment of matter of negligible dimensions, containing, nevertheless, an enormous number of elementary corpuscles. But the objection retains all its validity in atomic physics when, having assumed the existence of elementary corpuscles, we pretend to apply to these corpuscles the laws of classical mechanics of material points or even laws of a different form but implying the validity of the usual notions of space and time. Without dwelling further on this question which we shall have the opportunity to take up again, we shall close these few remarks on the dynamics of material systems.

4. Analytical Mechanics and the Theory of Jacobi

ANALYTICAL MECHANICS, with which the name of the great Lagrange is associated, is mainly a body of methods which permit us to quickly write the equations of motion of a system, when we know the variation of the variables, the knowledge of which is sufficient to fix the position of the system at each instant. Since we by no means wish to enter here into a detailed discussion of the methods of analytical mechanics, we shall limit ourselves to remarking that they converge to two well-known groups of equations: the equations of Lagrange and those of Hamilton. The method of Lagrange and that of Hamilton stand in opposition to each other in this respect, that in the method of Lagrange the energy of the system is defined with the aid of generalized velocities, i.e., derivatives of the parameters of position with respect to time, while in the method of Hamilton, the expression of the energy is a function of generalized momenta or "moments of Lagrange." In the framework of classical conceptions, we can moreover always pass very easily from generalized velocities to the moments of Lagrange and conversely, since momenta are always defined there with the help of velocities, so that the equations of Lagrange and those of Hamilton, in the case when either one can be effectively written, differ only in their form, and in the last analysis are equivalent. Now, when we come to quantum mechanics we shall see that the equations of Hamilton, suitably transposed, retain significance which is out of the question for the equations of Lagrange. This is easily understood if it is noted that in quantum theory dynamic notions retain their meaning while the sense of kinematic notions becomes obscured: momentum, which in classical ideas seems rather like a quantity derived from velocity, in quantum mechanics seems like a fundamental and autonomous quantity, independent of the concept of velocity, a concept whose significance is no longer well defined in all cases.

A very interesting and important chapter in analytical mechanics from the point of view we are considering is made up by the theory of Jacobi. Indeed this theory leads to a classification of the possible motions of a material point in a given field in a way that prepares the transition from the older mechanics to wave mechanics. We cannot develop here in detail the theory of Jacobi, which demands quite a complicated mathematical treatment, so we shall limit ourselves to a résumé of it by taking up the particular, but important, case of a permanent field of force, that is, one independent of the time. The ensemble of possible trajectories of a material point in a field of force depends on 6 parameters, since each of these trajectories depends on the initial position and the initial velocity of the material point. But it is possible to classify these trajectories into families depending only on three parameters, the trajectories of a certain family proving to be curves that are orthogonal to a certain family of surfaces. If one then succeeds in determining one of these families of surfaces, all the orthogonal curves are possible trajectories of the material point and the theory of Jacobi teaches us precisely how to determine the families of surfaces in question by starting with a certain partial differential equation of the first order and second degree which is named the Jacobi equation. To set up this equation, we start with the Hamiltonian expression of the energy, i.e., with the expression of the energy of the material point at each instant as a function of the values of the components of its momentum and its coordinates at that instant.

Hence we see that, thanks to the theory of Jacobi, we succeed in classifying the sextuple infinity of trajectories of the material point into families each containing a triple infinity of trajectories and each corresponding to a family of orthogonal surfaces. Each family of trajectories and the corresponding orthogonal surfaces are in exactly the same relation as the rays and wave fronts of a wave propagation as pictured in the manner of geometrical optics. The Scotch geometer, Hamilton, had, more than a century ago, noted this analogy and had drawn from it a very suggestive method of exposition of this aspect of analytical mechanics, but it is only the recent development of the quantum theory which has let us see in this analogy more than a simple similarity of mathematical form.

It is interesting to note in this connection that with the classical conception of a material point, the picture of a wave propagation as furnished by the theory of Jacobi can have only an abstract sense. In fact, in the classical conception, the material point, having at each instant a well-determined position and velocity, describes in the field of force a unique trajectory whose nature depends on the initial conditions. The family of trajectories grouped together by the theory of Jacobi are only possible trajectories and just one of them is actually realized in each case. Hence the families of trajectories have really a rather abstract significance since they represent an ensemble of possibilities of which only one, at most, is realized. Nevertheless, there might be a way of giving a concrete sense to the ensemble of trajectories considered by the theory of Jacobi: imagine that we have at our disposal an infinity of identical material points, exerting no influence on each other. There would then exist the possibility of supposing that the material points would describe the various families of trajectories which thus would turn out to be realized in a concrete fashion. In this way it can be seen that the theory of Jacobi is in a sense a statistical theory, since it pictures simultaneously ensembles of trajectories. Hence, we have an inkling that it contains in embryo the probabilistic and statistical interpretations of wave mechanics and we shall later see that this is very much so.

In the above lines, we have sketched the theory of Jacobi in the case of the motion of a material point in a given field. If it is desired to extend the same considerations to the case of an ensemble of interacting material points, a particular idea is introduced which later on we shall find again in the wave mechanics of systems: if the system is composed of N material points, it is necessary to imagine an abstract space formed by the 3N coordinates of the N members of the system, a space which is called the space of configuration. Accordingly, if the equation of Jacobi is set up for the system beginning with the Hamiltonian expression of the energy, an equation is obtained in partial derivatives of first order and second degree, involving the totality of the 3N coordinates of the points of the system and, consequently, this equation permits defining families of surfaces in the space of configuration (and no longer in the usual three-dimensional space). Now, it is evident that each configuration of the system is defined if the 3N coordinates of the constituents are given and can be represented geometrically by a point in the space of configuration: whence the name given to this space. A sequence of successive states of the system is therefore symbolized by a curve in the space of configuration: it is the trajectory of a representative point of the system. These symbolic trajectories of the system depend on 6N parameters, the six initial conditions relative to each of the N points, and the theory of Jacobi allows us to classify into families this 6N-tuple infinity of possible trajectories. Each of these families will depend on 3N parameters and will be formed by curves orthogonal to a family of integral surfaces of the equation of Jacobi. But this time it is in the space of configuration of 3N dimensions where we have the analog of a wave propagation. Hence we catch sight of the fact that in order to treat the problems of the dynamics of systems, wave mechanics, guided by the theory of Jacobi, will be obliged to follow it on this ground and consider the propagation of waves in the space of configuration: this will impose on the waves of wave mechanics not only a probabilistic and statistical significance to which we have already alluded, but also an abstract and symbolic character, very different from the kind of waves which classical physics pictures.

5. The Principle of Least Action

IT IS POSSIBLE to deduce the equations of the dynamics of a material point in a field of force derived from a potential by a principle which, in its general form, bears the name of the principle of Hamilton or of stationary action. According to this principle, the time integral, taken between the two limits t_1 *and* t_2, of the difference of the kinetic and potential energies of the material point is smaller (or larger) for the actual motion than for any other infinitesimally different motion which carries the material point from the same initial position to the same final position.

The principle of stationary action takes a particularly simple form in the important case of a permanent field. It then becomes the principle of Least Action of Maupertuis, according to which the path actually followed by the material point in going from a point A to a point B in a permanent field is the curve which makes minimum the line integral, "circulation of momentum" in relation to any other curve infinitely near joining the points A and B. The principle of Maupertuis can be derived from the principle of Hamilton, but it can also be associated with the theory of Jacobi. We have seen that in this theory, the trajectories in a permanent field could be considered as curves orthogonal to a certain family of surfaces: a simple reasoning allows us to deduce from this that these trajectories are determined by the condition of making minimum a certain integral and this integral proves to be the action of Maupertuis, i.e., the line integral of the momentum. This manner of proving the principle of Least Action is very interesting because it shows the relationship of this principle with the principle of minimum time of Fermat. We have indeed seen that the theory of Jacobi leads to treating these trajectories like the rays in a wave propagation conceived in the manner of geometrical optics. By examining in this light the reasoning which justified the principle of Least Action, it is seen that this reasoning is identical to that which was used in geometrical optics to justify the principle of minimum time or the principle of Fermat. Let us recall the statement of Fermat's principle: it asserts that in a refracting medium in a permanent state, the ray which passes through two fixed points A and B coincides with the curve which makes minimum the time taken by the light to go from A to B, i.e., it makes minimum the line integral of the reciprocal of the velocity of propagation. The relationship of the principle of Maupertuis to that of Fermat is thus evident. Nevertheless, between these two principles there remains an important difference: in the principle of Least Action it is the momentum which figures in the stationary integral in such a way that the latter has the physical dimensions of action (energy x time or momentum x length); in the principle of Fermat, on the contrary, it is the reciprocal of the velocity of propagation. For this reason it seemed impossible for a long time to consider the analogy of the two principles as anything other than a purely formal analogy having no deep physical basis. It even seemed that there was a marked opposition between the two principles from a physical point of view since, the momentum being proportional to the velocity, the integral of Maupertuis contains the velocity in the numerator while the integral of Fermat contains it in the denominator. This circumstance played a very important role at the time when the wave theory of light, due to the genius of Fresnel, succeeded in getting the better of the antithetical emission theory. By relying precisely on this different role played by the velocity in the expression for the integrals of Maupertuis and Fermat, it was thought possible to conclude that the famous experiment of Foucault and Fizeau, according to which the velocity of propagation of light in water is less than its velocity in a vacuum, carried an irrefutable and decisive argument in favor of the wave theory. But, it was assumed, not only in order to contrast the two principles of mechanics and geometrical optics but also in order to interpret the experiment of Foucault and Fizeau, that it was legitimate to treat the velocity of a material point figuring in the integral of Maupertuis like the velocity of propagation figuring in a different fashion in the integral of Fermat. Only wave mechanics, by showing that to the motion of any material point there must be associated a wave-propagation whose velocity of propagation varies inversely as the velocity of the material point, has put in their true light the profound relationship of the two great principles and the physical significance of this relationship. It has also shown that the experiment of Fizeau was not as crucial as was believed: this experiment proves quite well that the propagation of light must be represented as a propagation of waves and that the index of refraction must be defined by means of the velocity of propagation, but in itself, it does not completely exclude the existence of a granular structure of light if a suitable association of the waves and the grains of light is made. But these are questions of which we shall speak later on.

We have shown the analogy between the principle of Maupertuis and that of Fermat by comparing, in the main, the motion of a material point in a permanent field with the propagation of a wave in a refracting medium whose state does not depend on the time. By comparing the motion of a material point in a field which varies with the time with the propagation of a wave in a refracting medium whose state is progressively changing, we would succeed in showing the analogy between the principle of Least Action in its general form due to Hamilton and a generalized principle of Fermat applicable to non-permanent refracting media. We shall not dwell on this generalization: it will be sufficient for us to note that the fundamental analogy of the great principles of mechanics and geometrical optics is valid even beyond the very important, albeit special, case of permanent states.

Naturally, there is a principle of stationary action for systems of material points. But here, in order to make the wording precise, it is useful to consider the space of configuration, as previously defined, corresponding to the system. Let us limit ourselves, as an example, to the case where the potential energy of the system does not depend explicitly on the time; viz., the case of an isolated system which is not subjected to any external action, for then the potential energy results solely from the mutual actions and does not depend explicitly on the time. In this case, we have a principle of Least Action under the Maupertuisian form that we shall state with the help of a space of configuration of 3N dimensions and by considering in this space the vector whose 3N components are the components of the N momenta of the material points of the system. This principle of Least Action tells us that the path of a representative point of the system passing through two fixed points A and B of the space of configuration is such that the line integral (circulation) of the vector we have just mentioned, taken from A to B along the path, is minimum in relation to any other curve infinitely near in the space of configuration joining the points A and B. This principle is also easily proved by starting from the theory of Jacobi, and its analogy with the principle of Fermat still stems from the possibility of considering the same kind of paths of the representative point in the space of configuration as the rays of a certain propagation of waves in this space. Once again it is seen that, for systems, the transition from classical mechanics to wave mechanics will necessarily be effected in the abstract realm of the space of configuration.

Chapter 2 | Classical Physics

1. The Extensions of Mechanics

WE DID NOT have the intention of giving a complete sketch of classical mechanics in the few pages of the last chapter. It would be even less possible for us to give a complete overall view of classical physics in the present chapter. At most we can try to characterize its principle branches and make a few remarks about each of them.

One main branch of classical physics is formed by the various direct extensions of mechanics: hydrodynamics, the study of fluids, acoustics, the theory of elasticity. These sciences early claimed the attention of physicists because the phenomena they study force themselves on our attention in our daily life. From a theoretical point of view they appear to be immediate extensions of mechanics from which they borrow their fundamental principles and their methods of reasoning, augmented by a certain number of hypotheses suggested by experience. They do not explicitly introduce the idea that liquid, solid or gaseous bodies have an atomic structure, but on the contrary, they reason as if matter were continuous, isolating in this continuum, elements of volume for which they calculate the interaction with the neighboring elements of volume and to which they apply the laws of mechanics. Nevertheless, nothing stands in the way of reconciling these methods with the hypothesis of an atomic structure of matter if we suppose that the elements of volume about which we are thinking, although very small, are already large enough to contain an enormous number of molecules and to possess the properties of matter taken *en masse.*

These sciences, extensions of mechanics, although resting on principles which spring very simply from the laws of mechanics, are in reality difficult sciences that demand great effort and great ability on the part of the experimentalists and theoreticians. The physical data in these domains are complex and often difficult to study: the aid of higher mathematics is required to carry out the calculations. So, although in existence for a long time now, these sciences still have much further progress to make. They are essential in their applications for engineering work. In order to put themselves within the reach of practical men, who are preoccupied more with immediate results than with general theories, they have had to take approximate forms such as those called hydraulics or resistance of materials.

We shall not dwell any longer on these disciplines. The transformations of modern physics have modified them little and until now quanta have played an insignificant role in them. Hence they lie outside the main body of our study.

2. *Optics*

IF HYDRODYNAMICS and the theory of elasticity do not directly interest those who above all wish to study quanta, it is quite the contrary with optics whose progress is most intimately connected with the recent advances in physics. Just as in the case of the motions of solid and liquid bodies, the phenomena of light have always, and of necessity, attracted the attention of men. But it was only in the 17th century that optics became a true science. The laws of Descartes, which cover in a precise fashion the phenomena of reflection and refraction, and that principle of Fermat of which we have already spoken and which encompasses all of geometrical optics, were both stated at that time. In all this period in the history of optics the notion of the light ray played the fundamental role: one studied the rectilinear propagation of the rays in a vacuum or in homogeneous media, the bending of the rays at the surface of a mirror or upon entering a refracting medium, and the progressive curvature of the rays in a non-homogeneous refracting medium. It was at that time that Huyghens developed another way of interpreting the same phenomena by using the notion of waves and wave-fronts. In addition he showed that the recently discovered phenomenon of double refraction in Iceland spar could be interpreted by following this method. From a purely geometrical point of view there is an equivalence between the method which utilizes the consideration of rays and that which utilizes the consideration of wave-fronts. The propositions of geometrical optics let us see this equivalence and pass without difficulty from one point of view to the other. As we have already mentioned in the preceding chapter, the rays are curves orthogonal to families of wave-fronts and the principle of Fermat is an immediate consequence of this fact. But if there is a mathematical equivalence between the various ways of visualizing the problems of geometrical optics, one is led to very different conceptions of light according to whether one attributes the basic role to the rays or to the wave-surfaces. By considering the light ray as the essential notion, one is led to a corpuscular conception of light according to which the latter is formed of small corpuscles in rapid motion whose trajectories are the rays. The rectilinear form of the rays (rectilinear propagation) and the reflection of light at mirrors then find a very natural and intuitive explanation and refraction can also be interpreted. From this point of view, the rays have a physical significance, being trajectories of the light corpuscles, the wave-surfaces being only geometrical conceptions which permit a bundle of rays to be grouped into a single family, just as consideration of the integral surfaces of the Jacobi equation permits an ensemble of trajectories to be grouped into a family. But one can, on the contrary, consider the wave-surfaces as the essential reality and thus arrive at an undulatory conception of the nature of light. Light is then conceived as being constituted of waves which are propagated in space, the rays being only the abstractly defined curves orthogonal to successive wave-fronts. The keen analyses of Huyghens showed very well that this wave theory of light accounts for the pheonomena of reflection and refraction, but it is not easy to see at first how it can explain rectilinear propagation in homogeneous media, a physical fact whose interpretation seems so evident in the corpuscular hypothesis where it follows from the principle of inertia.

These two opposing concepts, the corpuscular concept or "theory of emission" and the wave concept shared the favor of the scholars of the 17th and 18th centuries. The former, first advanced by Descartes, found a defender of great authority in the person of Newton. The gifted creator of celestial mechanics had been struck by some difficulties which underlay the wave concept, notably in interpreting rectilinear propagation, and he spoke out clearly in favor of the corpuscular hypothesis. After Newton, the scientists of the 18th century were in general favorable to this way of representing light; and the wave concept, so brilliantly upheld at the end of the preceding century by Huyghens, did not then have more than a few isolated defenders (Euler). At that moment the fight might appear to have been won by the partisans of a discontinuous structure of light.

The beginning of the 19th century saw a complete reversal of this situation. The reason for this development was the discovery of the phenomena of interference and diffraction. A few special cases of these phenomena had been discovered at the time of Newton, first by Hooke and Grimaldi and later by Newton himself. That beautiful pheonomenon, known to this day by the name of Newton's rings, is actually an interference phenomenon. With his usual perspicacity Newton very clearly saw that an interpretation of these phenomena would require, even in the framework of the corpuscular theory to which he subscribed, the intervention of an element of periodicity. So he had imagined that the light corpuscles would alternately undergo a "fit" of easy transmission and a "fit" of easy reflection, a theory which at first approach might seem complicated and bizarre, but which in reality constitutes a first attempt at reconciling the corpuscular and undulatory concepts of light and, hence, two centuries in advance, heralds the theories of the present day. The 18th century, dominated by the idea that light is of a corpuscular nature, did not seem to have attached to interference phenomena all the attention they deserved. It is only at the end of the century and at the beginning of the following that the English physicist, Thomas Young, seriously resumed the study of these phenomena; but giving a complete and definitive explanation of them was reserved for the genius of a Frenchman, Augustin Fresnel (1788-1827). Returning to the wave concepts of Huyghens, Fresnel found in them a complete explanation of all the aspects of diffraction and interference known in his time; he succeeded in proving—this is an essential point—that the undulatory nature of light is not in contradiction with a rectilinear propagation in homogeneous media. Criticized by the adversaries of the wave theory because his interpretations lead to certain predictions of a paradoxical nature, he showed by experiment that these predictions are verified. From then on the triumph of his ideas was assured, and although still unheld at this time by scientists such as Biot and Laplace, the corpuscular concept went into a full decline and every day lost some of its supporters.

But Fresnel did not stop there. In order to account for the phenomena of polarization, he introduced the idea of the transversality of the light vibration which explains why polarized light has properties which are not isotropic at right angles to its direction of propagation. By studying the properties of transverse vibrations, Fresnel developed the theory of the intensity of reflection at the surface of a refracting body as well as that of the propagation of light in anisotropic media from which spring the existence and the laws of double refraction; and this whole body of doctrine, a veritable masterpiece of theoretical physics, is met with in all present day treatises on physical optics without important modification. Exhausted by this tremendous intellectual effort, Augustin Fresnel succumbed to an illness and died in 1827 at the age of 39, but the work which he had accomplished was praiseworthy and remains one of the finest chapters in the history of the development of physics.

After the death of Fresnel the wave nature of light was more and more generally admitted and the experiment of Foucault and Fizeau of which we have already spoken seemed to add an unanswerable proof in favor of this hypothesis. But, as we shall see, it is only much later, around the beginning of the present century, that the attention of physicists was redirected toward a corpuscular picture of light without their being permitted in doing this to think for an instant of abandoning the undulatory interpretations of Fresnel. It has consequently been necessary to attempt a sort of synthesis, or rather juxtaposition, of the corpuscular and undulatory pictures. Thus can it be seen that if Fresnel was correct in giving a general interpretation by means of waves for the phenomena of optics known in his time or discovered by him, the physicists of the opposite school were nevertheless not really wrong in suspecting the existence of a discontinuous aspect of light. The intuition of a Newton or of a Biot did not completely deceive them by making them suspect that the properties of light rays were closely related to those of the trajectories of material points in mechanics. It is not by accident that geometrical optics offers analogies with dynamics, that the principle of Fermat in particular is modeled on the principle of Least Action. The great theorems of analytical mechanics, above all the theory of Jacobi, explain the true meaning of the laws of geometrical optics, but the optics of waves in its turn indicates the path to be followed to broaden classical mechanics and teaches us then that classical mechanics, just as geometrical optics, is only an approximation, often valid, but whose domain of application is nonetheless limited.

We shall have to return to these questions, but perhaps, in order to pave the way, it will be useful to show for the time being in what way wave optics has been able to absorb geometrical optics, e.g., the principle of Fermat can be justified from Fresnel's point of view. The equation which represents the propagation of waves in the wave theory is an equation in second order partial derivatives, well-known by the name of the "Wave Equation." This equation contains a quantity, the phase velocity, which is a certain function of the coordinates of space and time in the most general case of propagation in a non-permanent refracting medium. In the important case of media in a permanent state, the velocity of propagation is independent of the time and defines at each point a constant "index of refraction." The equation of propagation then admits of monochromatic solutions which represent light of different frequencies, i.e., of different colors, which can be propagated in the given medium. It can be shown that if the index of refraction of the medium does not sensibly vary in the distance of the order of a wave length, the variations in the phase of the wave are represented quite closely by an equation in partial derivatives of the first order and second degree called "the equation of geometrical optics" whose form is exactly the same as that of the equation of Jacobi. The equation of geometrical optics permits us to find for each monochromatic wave propagation the families of surfaces, the wave-fronts, at which the phase has the same value: then the curves orthogonal to the wave-fronts can be found and they can be defined as the light rays corresponding to the propagation. From this the principle of Fermat, the theorem of Malus, the construction of Huyghens, and all the other laws of geometrical optics can be deduced. From the wave point of view, geometrical optics is valid whenever the rigorous wave equation can be replaced approximately by the equation of geometrical optics. The condition for this, as we have seen, is that the index of refraction does not vary too rapidly from point to point in space, but in addition it is necessary that there should be no obstacle in the path of the light hampering its free propagation and bringing about the appearance of the phenomena of interference and diffraction. Thus it is that in the eyes of a wave theorist, geometrical optics appears to be an approximation which is often valid but which has however only a limited domain of validity.

Let us return to the physical meaning of the wave theory. Since light waves traverse empty space without difficulty, it is not matter which transmits them. What then is the carrier of these waves, what medium is it whose vibration constitutes the light vibration? Such was the question that was asked of the protagonists of the wave theory. In order to answer this, they imagined a subtle medium, the luminiferous ether, distributed throughout the entire universe, filling all the empty spaces and impregnating material bodies. The properties of this mysterious medium had to be sufficient to explain the circumstances of the propagation of light *in vacuo:* the interaction of the ether and matter had to account for the propagation of light in refracting media. The followers of Fresnel devoted themselves to the problem of the ether: they sought to make its mechanical nature precise, to picture its structure. The results of this research were strange indeed: the ether, considered as an elastic medium, had to be a medium infinitely more rigid than steel since it can transmit only transverse vibrations, but this most rigid medium meanwhile offers no friction to bodies that move through it and in no way impedes the motions of the planets. No completely coherent theory of this paradoxical medium could be constructed and a great many physicists have come to doubt the real existence of this mental entity. We shall see further on how this question has later evolved, first in the electromagnetic theory, and then in the theory of relativity.

3. Electricity and the Electromagnetic Theory

IF MECHANICS and its extensions, acoustics and optics, are sciences whose origins are very old because they study phenomena, the existence of which men have been conscious for all time, the science of electricity, on the contrary, is of recent origin. Certainly certain facts such as the electrification of bodies by friction or the properties of natural magnets were known for a long time and the tremendous and awe-inspiring phenomenon of storms could never be passed unnoticed: but it is doubtful if these various facts had been sufficiently collated before the end of the 18th century so that anyone saw that here was the subject matter of an autonomous science forming a new branch of physics. It was really the end of the 18th and the beginning of the 19th century that brought this revelation. It is interesting to note that this was also the period of the discovery of interference and the development of wave optics. This marvellous period in the history of science, which saw the formation of modern optics and electricity, is for macroscopic physics the counterpart of what the last fifty years have been for atomic physics.

We do not wish to follow in detail the history of the development of electricity nor to analyze the contributions of Volta, Coulomb, Oersted, Davy, Biot, Laplace, Gauss, Ampère, Faraday and other physicists of the same period in building this new science. This study would be interesting, surely, but it would be long and would draw us too far from the subject we are considering. Therefore we shall limit ourselves to observing that after the middle of the 19th century the laws of electricity were well enough known, so that it was possible to attempt a synthesis of them and to seek to unite them into a body of homogeneous doctrine. It was this considerable work which John Clerk Maxwell, enlightened by the work of his predecessors and backed by great personal gifts, was able to accomplish in creating the general electromagnetic theory which bears his name. Maxwell succeeded in summarizing in a single system of equations, to which his name remains associated, all of the laws of electricity. The equations of Maxwell comprise two vectorial equations, representing six equations written between the components, and two scalar equations. On one side of these equations appear the components of the fields and of the electric and magnetic inductions; on the other side, the densities of the electric charges and currents. One of the vector equations expresses the great law of induction discovered by Faraday; one of the scalar equations expresses the fact that it is impossible to isolate a magnetic pole, while the other scalar translates Gauss' theorem of the flux of electric force. But, in writing down the second vector relation Maxwell brought to the theory his essential personal contribution. This second relation has as its object interpreting the manner in which the magnetic field is related to the current in accordance with the laws discovered by Ampère. We are led by these to write down that the curl of the magnetic field is equal (with a constant factor depending on the units) to the density of the electric current. But Maxwell saw that if the electric current figuring in these equations is defined solely as the flux of electricity, certain difficulties are met with, and to avoid them he had the admirable idea of completing the expression for the current by adding to the term which represents the displacement of electricity by conduction or convection, another term connected with the instantaneous variation of the electric induction. This new term represents a new kind of current, the displacement current, which is not necessarily connected with the motion of electricity: in polarizable media, a part of the displacement current can be interpreted by the motion of charges of free electricity which appear as a result of polarization, but another part of the displacement current, which always exists, even in a vacuum when the electric field varies there, is absolutely independent of the motion of electricity. Thanks to the introduction of the displacement current, the difficulties to which we made mention are removed and the difficult question of open and closed currents, which preoccupied the theorists of that time, was cleared up, for, if the displacement current is taken into account, there are none but closed currents.

But the really brilliant idea of Maxwell, after having written the general equations of electric phenomena, was to see in these equations the possibility of considering light as an electromagnetic disturbance. He thereby caused the whole science of optics to be included in the framework of electromagnetism, thus uniting two domains which had seemed entirely distinct and realizing one of the most beautiful syntheses of which the history of physics offers us an example.

In order to understand how Maxwell realized this synthesis, it is necessary to understand that the electromagnetic equations contain a constant which represents the ratio of the units of charge or of field in the system of electromagnetic units and in the system of electrostatic units. By combining the fundamental equations, it is easily shown that electromagnetic fields are propagated in a vacuum in accordance with the wave equation and with a phase velocity equal to the constant in question. Therefore, if one wishes with Maxwell to interpret light waves as being electromagnetic disturbances, one is led to predict that the velocity of propagation of light *in vacuo* ought to have the same value, generally represented by the letter c, as the ratio of these units. The numerical data known at the time of Maxwell permitted even then the assertion that this equality was exact to within 3 or 4 percent. All measurements made since then tend to indicate that this equality is exact. This fact brings a striking confirmation to the electromagnetic concept of light as proposed by Maxwell.

In Maxwell's ideas, a plane monochromatic light wave in a vacuum is characterized by two vectors, the electric and magnetic fields, vibrating with the frequency of the wave and propagating themselves in the direction of propagation; they are equal, perpendicular to each other and to the direction of propagation and they are in phase. All the results of Fresnel's theory can be found again by comparing the elastic vibration of the ether to this electric vibration: it is sufficient, one might say, to translate the reasoning into another language. In the electromagnetic theory, one can very well say no more about an ether: it is enough to consider the properties of empty space as being defined at each point by two vectors, the electric field and the magnetic field. The theory then takes on that abstract aspect which so often characterizes the theories of modern physics: it becomes in essence a system of equations. This abstract aspect of the electromagnetic theory is particularly evident in the form which Hertz gave to it sometime after Maxwell. Nevertheless, a great many physicists of this period still felt the need of giving a prop to the electromagnetic field, to consider it as a state of something. Great efforts were made, notably by Lord Kelvin, to obtain a mechanical representation of electromagnetic phenomena with the help of tensions and deformations in the ether, but these representations were never completely satisfactory and ended by falling into discredit. Since then the ether has served only as a hypothetical medium of reference which permits us to define systems of coordinates with reference to which the equations of Maxwell are valid in their ordinary form. Even when reduced to this modest role, the ether still proved troublesome: the electrodynamics of bodies in motion based on the idea that the ether can serve to define the axes of absolute rest, was a complicated doctrine which, at last, was shown not to be in accord with experiment. The theory of relativity has cleared up this situation by taking the lead in completely abandoning the notion of the ether.

One of the most striking confirmations of Maxwell's ideas has been the discovery by Hertz of electromagnetic waves which are called Hertzian oscillations. The electromagnetic theory predicts indeed that, if one succeeds in producing in an electric circuit electromagnetic phenomena of sufficiently high frequency, there can be an emission of an electromagnetic wave into the surrounding space which, according to Maxwell's ideas, should have an absolutely analogous structure to that of light waves. But the waves that a suitable electric circuit can be made to emit always have a frequency which is a great deal smaller, and a wave length a great deal longer, than those of light waves. From this fact there naturally arise differences in important properties: Hertzian waves do not act on our senses and because of their great length easily pass around extended obstacles. Nevertheless, despite these differences, there is a strong analogy between light waves and Hertzian waves and with the latter we can repeat the experiments of reflection, refraction, interference or diffraction which were classic experiments for the former; the experimental arrangements must, of course, be transposed to a much larger scale because of the increase in wave length. This memorable discovery of Hertzian waves and of their properties has left no doubt about the basic correctness of Maxwell's fundamental ideas on the subject of light. There is scarce need to recall that the discovery of Hertzian waves has led to radio and, later, to other types of telecommunications which derive from it.

Electromagnetic theory also allows us to study the propagation of light in material media. It leads to the well-known relation which connects the dialectric constant of a homogeneous substance with its index of refraction and it allows us to analyze the extinction of light in conducting media. But above all, when it is completed by the addition of the hypothesis of a discontinuous structure of the electricity contained in matter (electron hypothesis), it then permits us to analyze thoroughly the propagation of light in material media. We shall return to this idea in the next chapter.

4. *Thermodynamics*

We could not terminate this rapid examination of classical physics without saying a few words about a science which has been created in its entirety by scientists of the 19th century: thermodynamics. In the 18th century, it was assumed that heat is a fluid which is conserved, i.e., whose total quantity remains unchanged in the course of various physical transformations. This hypothesis is sufficient in a certain number of cases, notably in the study of the flow of heat in substances. The beautiful and classic theory of heat flow, due to Fourier, starts from relations which express the "conservation of caloric." But the many phenomena where heat appears through friction are interpreted with difficulty from this point of view and, little by little, physicists came to look on heat no longer as an indestructible substance, but as a form of energy. In all the purely mechanical phenomena which go on around us there always is as a matter of fact a conservation of energy except when friction with the appearance of heat is present. If heat can be considered as a form of energy, a general principle of the conservation of energy can be assumed. I do not have to recall here how this principle finally clearly appeared to the minds of the physicists around the middle of the last century or how it was confirmed by the measurement of the mechanical equivalent of heat. But, as is known, the principle of the conservation of energy is not enough to build a thermodynamics. It is necessary to add to it the principle of

Carnot or of the increase of entropy. This principle was first hinted at by Sadi Carnot, in 1824, when he wrote his reflections on the motive power of fire and found out that heat cannot wholly be transformed into work. From these reflections of Carnot there came, some years later, the principle that we use today. Clausius introduced the notion of entropy in order to express it and showed that the entropy of an isolated system always increases.

Relying on these two fundamental principles, there developed a thermodynamics which permits us to predict a considerable number of phenomena and plays an essential role especially in the theory of gases. It is an abstract science that occupies itself solely with energy stored in bodies and the quantities of heat or work exchanged between them. It does not try to enter into a detailed description of elementary phenomena, it is interested only in the larger aspects. So it is compatible with any number of different descriptions of the elementary phenomena: it lays down only the conditions which these descriptions must satisfy. Thus classical atomic physics, ignoring quanta, could give pictures of these phenomena in accord with the demands of thermodynamics, but quantum physics, although resting on very different conceptions, furnishes a picture equally compatible with thermodynamics. From the point of view of the constructive development of contemporary theories, thermodynamics has been able to serve as a guide in limiting the number of acceptable hypotheses, but without its indicating in a univocal fashion the way to be followed. Precisely because thermodynamics pictures only bulk appearances without wishing to enter into the details of the elementary processes, it does not run the risk of the errors which necessarily affect the more audacious theories that pretend to enter into a description of these processes. So, a great number of physicists were of the opinion, forty years ago, that it was preferable to be satisfied with these thermodynamic propositions without making an appeal to more precise, but more hazardous conceptions. This prudent method was named energetics. But if prudence is the mother of security, it is on the audacious that fortune smiles. So while the energeticists moved around on a solid but restricted terrain, the partisans of a more detailed description of elementary phenomena discovered new fields by developing atomistic and corpuscular conceptions. These conceptions have received so many confirmations from experiment, have brought to light so many hidden relations whose existence energetics could not even suspect, that today the old attitude of energetics corresponds to a stage that has been completely left behind. Continuing now our study of the development of classical physics, we must in turn delve into the new world of atoms and corpuscles.

Chapter 3 Atoms and Corpuscles

1. The Atomic Structure of Matter

It is well known that the thinkers of antiquity had a certain intuition of the atomic structure of matter. They had been led to this by a philosophic kind of idea that it is impossible to conceive of an indefinite divisibility of matter, that it is really necessary to stop somewhere in the operation that consists in considering smaller and smaller quantities of it. For them the atom was the ultimate element, indivisible, beyond which there was nothing more to look for. Modern physics, too, has arrived at an atomic conception of matter, but for it the atom is something very different from the atom of antiquity for it is a little structure, extended in space and of a complicated nature. In the conceptions of contemporary physicists, the true atoms (in the sense of the ancients) are the elementary corpuscles, for example electrons which today are considered (perhaps tentatively) as the ultimate constituents of atoms and hence of matter.

It is, as we know, the chemists who first introduced atoms in a precise fashion into modern science. The study of the properties of chemically well-defined substances, indeed, led the chemists to divide them into two classes: compound substances which can be broken down into simpler substances by appropriate operations and the simple substances which resist all attempts at decomposition.[1] These simple substances are also called elements. An examination of the quantitative laws according to which the elements unite in order to form compound substances had progressively led the chemists of the last century to adopt the following theory: an element is formed of small particles, all alike, called the atoms of this element; compound substances are made up of molecules formed by the union of several atoms of the elements. According to this hypothesis, to decompose a compound into its elements is to break up the molecules and liberate the atoms they contain. The list of elements, well-identified today, is already long. It contains 89 names and it is certain, for reasons to which we shall return, that if it were complete, it would contain at least 92 names. There are then at least 92 different kinds of atoms with which material bodies are built.

The atomic hypothesis has shown itself fruitful not only in explaining the fundamental facts of chemistry, but also in constructing physical theories. If bodies really are formed of atoms, it should indeed be possible to predict their physical properties by starting from this atomic structure. The well-known properties of gases, for example, ought to be interpretable by supposing them formed of a very great number of atoms or molecules in rapid motion. The pressure of a gas on the walls of the vessel which contains it should be due to the collisions of the molecules

[1] If at least we make exception of the transmutations realized by contemporary physicists.

against the walls. The temperature of the gas ought to be connected with the average agitation of the molecules, this agitation increasing as the temperature rises. This concept of gases has been developed under the name of the kinetic theory of gases and has led to a restatement of the laws of gases found by experiment. Besides, if the atomic concept is an exact representation of things, the properties of solid and liquid bodies should be interpretable by assuming that, for bodies in these physical states, the atoms or molecules are a great deal closer and in a greatly more constrained relationship than for bodies in a gaseous state. The considerable forces which are supposed to exist between atoms and molecules which are very close together should account for the properties of incompressibility, cohesion, etc., which characterize solid or liquid bodies. The theories developed along these lines met with some difficulties (several of which have since been cleared up by quantum theories), but the results have been generally satisfying enough to permit us to believe we are on the right road.

But, if the atomic hypothesis has shown itself fruitful as a basis for certain physical theories, it was no less indispensable for it, in order to be justified completely, to put its exactness in evidence by more or less direct experiments. This major work was accomplished, thirty years ago, by physicists at whose head it is necessary to place Jean Perrin whose experiments on this subject remain memorable. If it is impossible to perceive directly the motion of molecules or atoms, it is at least possible to note the random motions that these particles, by means of their continual collisions, impart to granules in suspension in a gas or liquid. The study of this motion of agitation, called "Brownian movements," permitted Jean Perrin to estimate the number of molecules contained in a gram-molecule of any gas at ordinary conditions of temperature and pressure. By virtue of a classic law of general chemistry due to Avogadro, it is known that this number is the same for all gases: it is called Avogadro's number. The experiments of Jean Perrin permitted him to assign to it a value between 6 and 7×10^{23} and all further experiments have confirmed this estimate in a remarkable degree. A great number of other more indirect methods can lead to an evaluation of Avogadro's number. These methods rest on the study of quite diverse phenomena: spectral distribution of radiant energy in thermodynamic equilibrium, scattering of light by a gas, emission of X-rays by radioactive substances. The evaluations thus obtained for Avogadro's number and the atomic quantities which are deduced from it (mass of the H atom, for example) show such an agreement that there can be no doubt of the correctness of the atomic hypothesis.

Thus the existence of atoms, postulated by the chemists, has been demonstrated by the physicists. It remains to be seen how it has been used by the theorists.

2. *The Kinetic Theory of Gases. Statistical Mechanics*

IF ONE ADOPTS the point of view which considers all bodies as being formed of atoms, one is led to think that, in bodies in the gaseous state, the atoms would be sufficiently separated from each other on the average so as to be for a good part of the time free from mutual interactions. From time to time, for an extremely short duration, one atom will be sufficiently close either to another atom of the gas, or to the wall of the enclosing vessel, for it to undergo an action: it is then said to undergo a collision with another atom or with the wall. Between two consecutive collisions, the atom will be displaced freely without being subjected to any sensible force and it can easily be seen that for all gases in their usual conditions, the total duration of the collisions experienced in a second by an atom is infinitely small compared to the total duration of the free distance, even though the number of collisions per second is enormous. If it is assumed that the laws of classical mechanics are applicable to atoms, the latter, between two collisions, ought to be endowed with a rectilinear and uniform motion and the collisions themselves, which give different results according to the manner in which they are produced, ought to satisfy the conservation of energy and momentum. If it is further admitted that the atoms, at least to predict the effect of the collisions, can be compared to rigid elastic spheres, the entire evolution of a gas should, in principle, be capable of calculation with the aid of the equations of classical mechanics. But if the picture of a gas formed of atoms or molecules compared to rigid elastic spheres is a well-defined problem, capable of a rigorous solution in principle, it is a problem of such complications that its rigorous and detailed solution is beyond all possibility; we can realize this by noting that there are under usual conditions in the order of 10^{19} atoms per cubic centimeter and that each of these atoms undergoes about 10^{10} collisions a second.

Hence, the problem might seem insoluble. And yet, the laws which govern gases are very simple, at least if we are content with first approximations (laws of perfect gases). It might seem then quite paradoxical that we can hope to account for these simple laws by starting out from a picture as complicated as that furnished by the atomic-kinetic conception of a gas. But it is just this extreme complexity of the picture that permits us to deduce the simple laws. Because of the extraordinarily great number of dynamic processes which are acting between the molecules of a gas, we are permitted to study the sum total of these processes with the aid of the calculus of probabilities and obtain laws of the mean with great exactness and often great simplicity. The possibility of observing an exception to these laws is extremely improbable by reason of the extraordinarily great number of elementary processes which go into forming the mean result.

The kinetic theory of gases was developed around the beginning of the second half of the 19th century, chiefly by Maxwell and Clausius: it has been, as one might say, codified by the work of Boltzmann. We do not intend to summarize the principal results obtained by this theory; they are well known today to all who have studied even a little theoretical physics. Let us recall only that the pressure exerted by a gas on the walls of the enclosing vessel has been interpreted as being due to the innumerable impacts of the gaseous molecules against the wall, that the temperature is considered as giving a measure of the mean kinetic energy of the molecules, that the equation of the state of a perfect gas is obtained very easily, that all sorts of interesting and exact predictions, to a first approximation, are obtained relative to the specific heats, to the diffusion of gases, to their viscosity, etc. Certainly many questions still remain to be answered in this domain and just recently the work of those such as Yves Rocard has opened new paths, but altogether it can be said that the kinetic conception of gases founded on the hypothesis of the atomic constitution of matter gives a good representation of reality.

One of the great successes of the kinetic theory of gases has been an interpretation of the notion of entropy. By analyzing the mutual impacts of the atoms of a gas and the establishment of a state of equilibrium by these impacts, Boltzmann was able to define a quantity which must always increase as a result of these impacts up to the moment when it attains a maximum characteristic of the state of equilibrium. This quantity must evidently be likened to entropy and Boltzmann has shown that it is equal to the logarithm of the probability of the instantaneous state of the gaseous mass. This remark has thrown a vivid light on the physical meaning of the notion of entropy which Henri Poincaré declared "prodigiously abstract." The theorem of Clausius, according to which the entropy of an isolated system always continually increases, signifies then that an isolated system always evolves spontaneously toward states which are more probable. This beautiful interpretation of entropy has constituted a signal success for the partisans of the atomic theory. While the energeticists had to assume the principle of entropy as an inexplicable, experimental fact, the kinetic theory of gases succeeded at once in understanding it by considering the statistical evolution of an immense number of atoms in random motion.

The kinetic theory of gases thus led the theorists to consider the over-all and statistical aspects of an enormous number of elementary and uncoordinated mechanical processes. It suggested then a systematic study of these aspects, relying both on the general laws of mechanics and on the principles of the calculus of probabilities. This study was indeed made by Boltzmann, and then by Gibbs, and has resulted in the creation of a new science, statistical mechanics. Not only does statistical mechanics let us re-establish all the essential results of the kinetic theory, but it also lets us set forth general results which are capable of being applied to aggregates of atoms or molecules other than gases, for example, solid bodies. An example of such is the celebrated theorem of the equipartition of energy, according to which, in a system formed by a great number of component parts and maintained at an absolute temperature T, the energy is divided between the different degrees of freedom of the system in such a way that each of these degrees of freedom possesses on the average the same quantity of energy proportional to T. Applied to gases, this theorem gives very interesting and often well-verified results; applied to solids, it leads to a prediction that the atomic heat of solid bodies ought to be, in general, equal to 6 (law of Dulong and Petit) and in any case, never lower than 3, and these predictions prove to be equally well-verified in a great number of cases. Nevertheless, if these rigorous predictions of statistical mechanics have often successfully undergone the test of experiment, they have proven at times insufficient: at low temperatures, the specific heat of gases (at constant volume) no longer varies as the theory predicts, and certain solid bodies (diamond) have an atomic heat much less than 3. These discordant facts were disturbing, for the methods of statistical mechanics are so general that they should not be liable to exceptions, and it was incomprehensible that alongside so many very well-verified predictions, the theory should suffer indisputable setbacks in certain cases. As we shall later see, it is the discovery of quanta which will clear up this situation by teaching us to recognize the limits of the validity of the methods of classical mechanics and consequently of the statistical mechanics of Gibbs and Boltzmann.

According to the interpretation which statistical mechanics gives to the results of thermodynamics, the laws of thermodynamics no longer have a character of rigorous necessity; they have only an extraordinarily high probability of being verified. Thus, for a gas contained in a given vessel and maintained at a constant temperature, the pressure or entropy of the gas as calculated by thermodynamic methods represents only the most probable values of these quantities, compatible with the imposed conditions, but these most probable values are so much more probable than any closely neighboring values that in practice they alone can be observed. Nevertheless, theoretically there are possible fluctuations of these instantaneous quantities in relation to their most probable values as calculated by thermodynamics. Most often these fluctuations are too small or too infrequent to be observed, but in certain favorable cases, however, they can become evident. One knows, for example, that the fluctuations of density in a gas in the vicinity of its critical point gives rise to observable manifestations (critical opalescence).

The success of statistical mechanics has accustomed physicists to consider certain laws of nature as having a statistical origin. Because there is an enormous number of elementary processes in a gaseous mass the pressure or entropy of the gas obeys simple laws. The laws of thermodynamics appear as laws of probability, statistical resultants of phenomena on the atomic scale which are impossible to study directly or to analyze in detail. The rigorous dynamical laws, the absolute determinism of mechanical phenomena are relegated to the atomic world where they become unobservable, and it is only their average probable consequences which are observable on a large scale. So attention was drawn for the first time to the importance of the laws of probability and to the fact that, in a great number of phenomena at least, the observable laws are the laws of averages. We shall see that wave mechanics has tended to accentuate this orientation and to assume that the observable laws for the elementary particles themselves are laws of probabilities.

3. *The Granular Structure of Electricity: Electrons and Protons*

By what we have just said, it is seen that in physics as well as in chemistry, the hypothesis, according to which all bodies are made up of molecules, which in turn are made up of various arrangements of elementary atoms, has proven fruitful and has received good experimental confirmation. But the physicists did not stop there: they wanted to know how the atoms themselves were constructed, and to understand how the atoms of various elements are differentiated from each other. They have been helped in this difficult task by the progress of our knowledge of the structure of electricity.

From the beginning of the study of electrical phenomena, it seemed natural to think of electricity as a fluid, for example, to consider the electric current which passes through a metallic wire as the flow of an electric fluid along this wire. But, as was known for a long time, there are two kinds of electricity: positive electricity and negative electricity. We must therefore suppose that there exist two different electric fluids, a positive fluid and a negative one. Now these fluids can be pictured in two different ways: they can be imagined as being formed by a substance occupying, in a continuous fashion, the entire region where the fluid is found, or else they can be represented as being formed by clouds of small corpuscles, each corpuscle like a little ball of electricity. Experiment has decided in favor of the second of these concepts: forty years ago we learned that *negative* electricity is formed of little corpuscles, all identical, with an extraordinarily small mass and electric charge. These corpuscles of negative electricity are called "electrons." We know that electrons were first observed in a free state outside of matter in the form of cathode rays produced in discharge tubes. And later we learned how to produce them by the photoelectric effect and by thermo-ionic emission from incandescent bodies. The discovery of radioactive substances next furnished new sources of electrons, for several of these substances spontaneously emit β-rays which are nothing else than electrons, generally with very high velocities. We have been able to verify that all electrons, whatever their manner of production, always carry the same extremely small negative charge. By studying their behavior when they move in empty space, we have been able to verify that they move as they should in conformity with the laws of mechanics of small electrified particles, and by observing how these small particles move in an electric or magnetic field, we have been able to measure their mass and electric charge, both extremely small quantities.

For positive electricity it has taken a little longer to obtain proof of its corpuscular structure. Nevertheless, the physicists have come to the conclusion that positive electricity in the last analysis could be considered as being formed of corpuscles, all identical, the protons. The proton has a mass which, although still very small, is almost two thousand times greater than that of the electron; this fact establishes a curious dissymmetry between positive and negative electricity. For the charge of a proton is equal, on the contrary, to that of an electron in absolute value, but, of course, positive instead of negative. Until recent times, the proton had been considered as being the elementary unit of positive electricity. But the discovery of the positive electron has come along to complicate matters: we shall later see, as a matter of fact, that we have been able to detect the existence of corpuscles of positive electricity having the same mass as the electron but with an electric charge equal and of opposite sign; these are the positive electrons or positrons. What then is the real elementary corpuscle of positive electricity? Is it the proton or positron? Or else should it be assumed that there are two elementary corpuscles of positive electricity, irreducible to each other? The discovery of neutrons, which preceded that of positive electrons by a short time, might lead one to believe that protons are complex, formed by the union of a neutron and a positive electron. Today we rather assume that the proton and the neutron are two states of the same elementary particle. Be that as it may, physicists have always assumed until recently, that the proton was the unit of positive electricity. It is this point of view that we wish to maintain here for the moment.

Electrons and protons have an extremely small mass, but nevertheless, their mass is not zero and a great number of electrons and protons will be able to exhibit a significant mass in total. It was therefore tempting to suppose that all material bodies, characterized essentially by the fact of having weight and inertia, that is by their mass, are formed solely, in the last analysis, of protons and electrons in enormous numbers. From this point of view, the atoms of the elements, which are the ultimate materials from which all material bodies are built, should in themselves be formed of protons and electrons and the 92 different kinds of atoms of the 92 elements should be made up of 92 different combinations of electrons and protons.

The question is then raised to find out what these combinations of electrons and protons could be, i.e., to construct "models of atoms." Various hypotheses have been proposed. One model proposed by Sir J. J. Thompson, the illustrious physicist, who has contributed so much by his efforts toward making our understanding of the constitution of matter precise, has had a certain vogue: it represented the atom as a homogeneous ball of positive electricity in which the negative electrons were found in equilibrium. But there is another model which has ended in brushing that one aside: this model, called today the Rutherford-Bohr model, represents the atom as a little solar system in miniature, formed of a central positive charge around which the electrons gravitate. Suggested first by Jean Perrin, this picture of the atom was confirmed by the study of the deflection of particles by matter. This study, made chiefly by Lord Rutherford and his collaborators, has shown that the positive charge of the atom was concentrated in a very small space at the center of the atom in accord with the planetary model. The central nucleus of the atom would thus carry a positive charge of electricity and would be surrounded by electrons, playing the role of planets and gravitating around it under the action of the Coulomb force. Each kind of atom will be characterized by the number N of the planetary electrons that it contains in its normal state. Since an atom in its normal state must be electrically neutral, we see that the atom with N planetary electrons must have a nucleus whose electric charge is equal and of opposite sign to that of the N electrons. In an atom with a solitary planetary electron, the nucleus must carry an equal and opposite charge to that of the electron, and all the other nuclei will carry positive charges which are multiples of the latter. The nucleus of an atom with one electron (hydrogen atom) can thus be considered as the unit of positive electricity: and this is precisely the proton of which we have already spoken. Each kind of atom is in this way characterized by a whole number N which is called its "atomic number," and it is then possible to arrange the 92 elements in a linear series along which the atomic number increases from 1 to 92. *A priori* it is probable that the classification thus obtained will sensibly coincide with the classification achieved by increasing atomic weights, for the more complex the nucleus becomes, the more its weight should increase. Several phenomena have permitted us to attribute with certainty a definite atomic number to the different elements: such a one is the displacement in the scale of frequencies of the homologous lines in the X-ray spectrum of the elements, a displacement which, according to an experimental law due to Mosely (1913), is proportional to the square of the atomic number. But for a few inversions, the series of increasing atomic numbers coincides very well with that of increasing atomic weights.

The planetary theory of the atom is thus found to be confirmed by experiment. In 1913, Bohr, in a celebrated memoir, succeeded in giving this theory a mathematical development which permits the exact prediction of Röntgen and optical spectra. But in order to obtain these remarkable results, Bohr was obliged to apply to the planetary model of the atom the guiding ideas of the quantum theory, for the application of the ideas of classical mechanics and electromagnetism to this model was unable to lead to any good result, as we shall explain later. Since Bohr's theory can be developed only with the aid of the quantum theory, we shall postpone its study until a later chapter.

4. Radiation

WE HAVE just briefly shown how modern physics, principally in the period between 1870 and 1910, had been able to advance our knowledge concerning the structure of matter and electricity. It is proper to say here a few words concerning the fashion in which it has advanced our knowledge of radiation.

The domain of optics and the wave theory was indeed considerably enlarged by the discovery of new kinds of waves which differ from ordinary light only by a greater or smaller wave length. These new waves had remained unknown for a long time because they do not impress our eye, but they are capable of exhibiting certain physical actions such as heating, photographic impressions, electrical effects, etc., and through these the physicists established their existence. To these waves, which, except for wave length, are identical in their nature with light, the generic name of radiation has been given and the different kinds of visible light appear to be no more than a small group in the vast family of radiation.

Thanks to the discoveries made in the last fifty years, we know today, without any gaps, all the radiations whose wave lengths extend from 50 km. to one ten-billionth of a mm. From 50 km. to 1/10 of a mm. extends the immense domain of Hertzian radiations, well-known because of their use in radio. From 1/10 mm. to 8/10,000 of a mm. we find the infra-red radiations with their strong heating effect; from 8/10,000 mm. to 4/10,000 mm. is found the ultra-violet with its strong chemical and photographic action. Then comes the large domain of Röntgen rays or X-rays from 1/10,000 mm. to about one one-hundred-millionth of a mm. And finally beyond that, with even shorter wave lengths, are found the very penetrating waves emitted by radio-active substances or γ-rays.

It is not necessary for us to recall here how all the radiations in this immense scale were successively discovered by a long series of admirable experiments. What is essential to note is that for all these radiations, the undulatory conception, which had been so brilliantly confirmed by the facts in the domain of visible light, was shown to be equally valid. With Hertzian waves, with X-rays and even with γ-rays, we have been able to obtain phenomena of a clearly undulatory character (refraction, interference, diffraction, diffusion). Thus there can no longer be any doubt today that for all radiation, the wave theory is valid in the same degree as for light. The different radiations differ from each other only in their wave lengths, and it is this variation which is the cause of all the differences in their properties. But we should remark for the present that, if the wave conception applies to all radiations, it has also met for all of them the same limitations in the course of the contemporary development of physics, and we shall see that the return to the corpuscular conception, expressed by the concept of the photon, has shown itself necessary in the whole domain of radiation. And this last remark completely shows that all radiations have essentially the same physical nature.

The discovery and classification of the different radiations, the proof of the identity of their nature, permitted scientists forty years ago to distinguish two distinct entities in the physical world: on one hand, matter formed of atoms, which themselves are constituted by collections of protons and electrons, i.e., the elementary grains of electricity; on the other, radiation made up of the whole scale of radiations which are identical in nature and differentiated from each other only by the value of their wave length. Matter and radiation are completely independent realities, for matter can exist without any radiation and radiation can traverse regions of space entirely empty of matter. Nevertheless, one of the essential problems of physics is the study of the interactions which occur between matter and radiation when the two exist together. It is necessary to try to analyze the actions of radiation on matter and the reaction of matter on radiation; it is necessary to try to understand the manner in which matter is capable of absorbing or emitting radiation. The first theory, which in modern physics tried to resolve these problems in a complete and detailed fashion, is the electron theory. We must now say a few words about it.

5. *The Electron Theory*

THE ELECTROMAGNETIC THEORY of Maxwell furnished equations representing exactly, on our scale, the relation between the measurable electromagnetic fields on one hand, and electric charges and currents on the other. Obtained by uniting into a single formal system the result of macroscopic experiments, their value was incontestable in this domain. But in order to describe the details of the electrical phenomena in the heart of matter and in the interior of atoms, to predict the radiation emitted or absorbed by these ultimate material particles, it was necessary to extrapolate Maxwell's equations and give them a form applicable to the study of phenomena on the atomic and corpuscular scale. That is what was done, with more audacity than would seem at first, by one of the great pioneers of modern theoretical physics, H. A. Lorentz.

Lorentz took as his point of departure the idea of introducing into the equations of electromagnetism the discontinuous structure of electricity. He assumed the existence of elementary corpuscles of electricity to which he gave the generic name of electrons and supposed that all matter is formed by combinations of these corpuscles. What we popularly call an electrically charged body is one which contains in total more corpuscles carrying electricity of one sign than corpuscles carrying electricity of the other sign. An electrically neutral body is a body containing in equal quantity corpuscles carrying each of the two kinds of electricity. Naturally in all material bodies on our level of experience, the number of electric corpuscles present is always enormous. From this point of view, the electric current which flows through a conductor is due to a displacement of all the electrons contained in the conductor so that the property of being a conductor is explained by a certain freedom of movement of the electrons in the conductor. On the contrary, the properties of insulators are explained by the fact that the electrons contained in the insulator have a position of equilibrium and can be moved from it only a small amount. Each electron creates around itself an electromagnetic field and the fields that we can observe and measure in our experiments are the statistical resultant of the superimposition of an enormous number of elementary fields due to the different electrons of matter. As often happens, this statistical resultant obeys simple laws, and these laws are the laws of Maxwell's theory which connects macroscopic fields with charges and currents that are directly observable. More bold than Maxwell's theory, Lorentz's theory seeks to describe microscopic electromagnetic phenomena from which the phenomena observable in our experiments result through an effect of the average. It seeks then to define electromagnetic fields, charges and currents at each point and at each instant, both in the spaces that separate electrons and in the very interior of the electrons. Lorentz assumed that the microscopic quantities, fields, charges, currents obey equations of the same form as the macroscopical equations of Maxwell, with this exception however, that it is no longer advisable to distinguish fields and the corresponding inductions and that the charges and currents are here to be expressed as a function of the very structure of electricity. It can be shown that by taking the average of the elementary microscopic phenomena, we can pass from the equations of Lorentz to those of Maxwell and at the same time interpret the difference between fields and inductions. The electromagnetism of Maxwell thus seems like a "gross" electromagnetism resulting by averaging the "fine" electromagnetism of Lorentz.

The electron theory, built on the basis we have just outlined, led to important successes in predicting a great number of phenomena. First, it permitted us to recover an interpretation of the laws of dispersion that had already been obtained by certain preceding theories. Then, it has, and this no doubt has been its most important success, permitted us to predict exactly the normal Zeeman effect, i.e., the way in which the spectral lines emitted by an atom are affected in the simplest case by the presence of a uniform magnetic field. The discovery of this effect of a magnetic field on the frequency of spectral lines has given a complete verification to the electron theory and by the change in frequency it can be shown that the particles in motion to which the spectral emission is connected are negative electrons whose normal existence in the interior of matter is thus demonstrated. The theory of Lorentz has won here a great success. In a general manner, it has also supplied satisfactory interpretations for all the phenomena where the action of an electric or magnetic field produces a variation in the normal conditions of emission, propagation or absorption of light. Such a case, for example, is the magnetic phenomenon of circular polarization (Faraday effect) which, in the light of the Lorentz theory, appears as an inverse Zeeman effect. Still other such phenomena are those of electrical and magnetic birefringence. In all the domain of electro-optics and magneto-optics, the services rendered by the Lorentz theory have been considerable.

The electron theory also seemed to bring a solution to an important problem: the origin of the emission of radiation by matter. According to the equations of Lorentz, an electron endowed with a rectilinear and uniform motion carries along bodily its electromagnetic field and, consequently there is in this case no emission of energy into the neighboring space. But if the motion of an electron undergoes an acceleration, it can be shown that there is an emission of an electromagnetic wave and the energy thus lost at each instant by the electron is proportional to the square of its acceleration. Since an alternating current is made up of innumerable electrons in periodic motion, it can at once be explained why such a current can radiate energy and thus we again find an explanation for the emission of Hertzian waves by alternating currents in an open circuit such as the currents which are set up in a radio antenna. We therefore again find the theory of the emission of Hertzian waves such as it is contained in the equations of Maxwell. But, in calculating the wave emitted by the accelerated motion of a single electron, the electron theory gives a "fine" image of this phenomenon of the emission of radiation by matter and therefore it should be possible in principle to explain the rise of electromagnetic waves on the atomic scale and show, for example, how the spectrum emitted by an atom arises from the motion of the electron contained in this atom. We shall see in a moment what difficulties the realization of this program has come up against. But at the beginning, the theory of the "acceleration wave" seemed to furnish a complete explanation of the emission of radiation by matter, and the fact that X-rays appear when an electron, striking a solid anticathode, undergoes a quick stoppage, seemed a very strong proof in favor of this opinion.

Despite its brilliant beginnings, the electron theory has not been sufficient to interpret the properties of matter on the atomic scale. We shall see that in studying the thermodynamic equilibrium between matter and radiation by means of Lorentz's equations, difficulties have arisen which could be removed only by the introduction of the entirely new concepts of the quantum theory. Then again, if we wish to interpret the radiation of atoms by the motion of intra-atomic electrons, it would be necessary to suppose that, in the normal state, the electrons within an atom are motionless; otherwise, obliged to move within the very small region of the atom, they would necessarily be endowed with highly accelerated motions and would constantly emit energy in the form of radiation, and this would be contrary to the very idea of the stability of the atom. Now, as we have seen, the progress of our knowledge of the atom has led us to adopt for the latter a planetary model with a continuous motion of the planetary electrons. Hence, there is a flagrant contradiction between the existence of stable states for the atom and the theory of the acceleration wave. Here, too, the solution has been obtained (by the theory of Bohr) only by introducing quantum concepts.

Thus we see by these few examples, which could be multiplied, that the electromagnetic theory, completed and extended by Lorentz by taking into account the discontinuity of electricity, has been able to explain brilliantly a number of phenomena but that in the atomic domain it has come to an impasse against the impossibility of understanding the experimental facts without making an appeal to ideas completely different from the so-called "classical" ideas on which it rests.

Chapter 4 The Theory of Relativity

1. The Principle of Relativity

BEFORE BEGINNING the study of the development of our knowledge about quanta, it is impossible not to devote a short chapter to the theory of relativity. Relativity and quanta are the two pillars of contemporary theoretical physics and, although we should like to direct our attention in this book chiefly to the second one, we can not pass over the first one entirely in silence.

The development of the relativity theory found its point of departure in the study of certain facts in the domain of the optics of bodies in motion. The Fresnel conception of light, as we have seen, assumed the existence of an ether, filling all the universe and penetrating all bodies, and which served as the vehicle for light waves. Maxwell's theory had modified the importance of the ether a little, for it no longer demanded that light waves be a vibration of anything; in it the light wave can be supposed to be defined uniquely by electromagnetic vectors. Since the attempts which were made to interpret mechanically the laws of electromagnetism had given little satisfaction, the fields of the Maxwell theory were finally considered as the primary data for which it was useless to seek an interpretation by mechanical images. From then on, the electromagnetic theory had no more need for invoking the existence of a vibrating elastic medium. The notion of an ether would seem to have become useless for the successors of Maxwell. In reality, it was nothing of the sort, and the successors of Maxwell, Lorentz in particular, have had to continue to speak of it. Why was this so? Because Maxwell's equations of electromagnetism do not satisfy the principle of mechanical relativity, i.e., if they are valid in a certain system of reference, they are no longer valid in a system endowed with a motion of rectilinear and uniform translation with respect to the first—at least if it is assumed, which seems self-evident, that in passing from the first system to the second, a transformation of coordinates must be made as it has always been made in classical mechanics. In classical mechanics, indeed, it was assumed that there exists an absolute time, valid for all observers, in all systems of reference, and in addition it was also assumed that the distance between two points in space has an equally absolute character, and should have the same value in all systems that can be used to fix the position of the points in space. These two principles, that seemed so natural to assume, immediately furnished the simple, classical formulae which give the transformation of coordinates when we pass from one system of reference to another which is in rectilinear, uniform, translatory motion with respect to the first. These formulae defined the Galilean transformation. Now, it is a fundamental theorem of classical mechanics that the equations of mechanics are invariant for the Galilean transformation. Valid in systems of reference connected with the system of fixed stars, the equations of Newton remain valid in any system in rectilinear uniform translatory motion with respect to the fixed stars, if the validity of the Galilean transformation is assumed for the transition from one of these systems to the other. But, on the contrary, the equations of the theories of Maxwell and Lorentz, very different in form from the equations of classical mechanics, are not invariant under a Galilean transformation. From this it must be concluded that, if the equations of Maxwell are valid in a certain system of reference, they are no longer so in a second system in rectilinear uniform motion with respect to the first. Everything acts then as if there existed in the universe a certain "medium" of reference, so that the equations of electromagnetism are valid for systems of axes fixed in this medium and for these alone. It is this medium of reference which the successors of Maxwell called by definition the ether. For them, then, the ether was no longer an elastic medium having a certain substantiality and capable of propagating light waves: it was no more than an abstract and conventional medium serving to set apart the systems of reference with respect to which the equations of Maxwell were valid.[1]

Now, even reduced to this modest role, the concept of the ether, as we have already shown, still proved troubling enough. Indeed, the observed phenomena of light, according to the Maxwell theory, should be influenced by the motion of the observer with respect to the ether and a physicist should be able, by making observations on the propagation of light, to deduce from them the velocity which he possesses with respect to the ether, and this would give a certain degree of substantiality to this mysterious entity. Now, to be precise, the terrestrial physicist,

[1] It is curious to note that Dirac has recently found it necessary to revive the ether concept in connection with his quantum theory of electrodynamics. He finds that with any point of space-time, even if devoid of matter or charge, there must be associated a velocity and this must be regarded as the velocity of some real physical thing (the ether). Tr.

who is experimenting in his laboratory, is always carried along with a great velocity by the motion of the earth around the sun and, besides, this velocity changes completely in direction from one time of the year to another, since the motion of the earth around the sun is approximately circular. Therefore, if by an improbable chance, the terrestrial physicist finds himself at a certain time of the year to be at rest with respect to the ether, several weeks or several months later he would be in rapid motion with respect to it. So, by experiments spaced throughout the length of the year, it must necessarily be possible to detect the motion of the earth with respect to the ether. But all the experiments, no matter how precise or how varied, made in optics by the scientists of the 19th century, could not detect the influence of the motion of the earth with respect to the ether. For a long time, nevertheless, this absence of results could be reconciled with the accepted theories because they predicted only extraordinarily small effects, less than those which the optical experiments, even though extremely precise, could detect. It can, as a matter of fact, be shown, that the motion of the observer with respect to the ether should cause only effects proportional to the square of the ratio of the velocity of the observer in the ether to the velocity of light *in vacuo*. As this ratio is always extremely small, the expected effects are extremely feeble. But, due to the incessant progress of experimental techniques, there came a moment when the experimenters were capable of detecting, by interference experiments, quantities of the order of those that the theory predicted as being attributable to the influence of the motion of the observer with respect to the ether. And yet, experiment again furnished a negative result: the expected effects, assuredly very small, but which would have been measurable as the theory predicted, could not be detected. The ether persisted in remaining elusive, resulting this time in a flagrant contradiction with the classical theory. That, then, was the far-reaching conclusion which, in 1881, came out of the celebrated experiment of Michelson, repeated a little later by Michelson and Morley. And other experiments which, too, should have detected the motion of the earth with respect to the ether by electromagnetic rather than by optical effects (experiment of Trouton and Noble, for example), had no more success than that of Michelson.

Numerous attempts, naturally, were made to try to reconcile the negative result of the Michelson experiment with existing theories. Fitzgerald and Lorentz proposed, in particular, to assume the existence of a contraction of material bodies brought about by their motion through the ether: this contraction should diminish the length of bodies in the direction of their motion while leaving unchanged their transverse dimensions, and should result in exactly compensating for the effects of the motion on the propagation of light. But this ingenious hypothesis obviously had a completely artificial character and seemed to be imagined solely to cover up a failure. It was, as we know, Albert Einstein who, by an admirable intellectual effort, was able to find the true solution to the problem. (1905)

The underlying reason that the electromagnetic theory of Maxwell and Lorentz seemed to imply the possibility of detecting the motion of uniform translation of an observer, with respect to the ether, by measurements in optics or electromagnetics, was that it was assumed *a priori* that, in passing from one system of coordinates to another system, endowed, with respect to the first, with a motion of rectilinear and uniform translation, the coordinates of the two systems are related by the formulae of the Galilean transformation. Now the Maxwell-Lorentz equations are not invariant for the Galilean transformation and hence there results, as we have seen, the possibility, not confirmed by facts, of detecting the motion of the earth with respect to the ether. But a mathematical study of the equations of electromagnetism had shown Lorentz that if these equations are not invariant for the Galilean transformation, they are so for another linear transformation of the coordinates, a little more complicated than the Galilean, and which is called today the Lorentz transformation. At first, this had seemed to be a simple mathematical curiosity; the Lorentz transformation did not seem to have a very clear physical significance. But it was one of the aspects of Einstein's brilliant idea to assume that the Lorentz transformation represents the true relation which physically exists between the coordinates used by two observers in uniform translation with respect to each other (if at least these two observers are also in uniform translation with respect to the system of fixed stars). Therefore it would not be the Galilean transformation, but rather that of Lorentz which would be physically exact in this case. Since the equations of electromagnetism are invariant for a Lorentz transformation, it then follows from this that the equations have the same form for two observers in relative uniform motion: all optical and electromagnetic phenomena are then rigorously the same for the two observers, and it is impossible that one of these phenomena would let one of the observers discover his absolute motion with respect to the ether. The negative result of the Michelson experiment and of other experiments designed to measure the motion of the earth with respect to the ether then becomes entirely natural. Inversely, if the "relativity" of all optical and electromagnetic phenomena is postulated as a principle in the same sense in which classical mechanics assumes the relativity of mechanical phenomena, it becomes necessary to assume that it is the Lorentz transformation and not the Galilean one which correctly expresses the relations between the coordinates of two observers in rectilinear and uniform relative motion.

It was vital to investigate the causes and the physical consequences of the necessity of substituting the Lorentz transformation for the Galilean. That is what Einstein did with a fine and profound critique of space and time. This critique was necessary, for the adoption of the Lorentz transformation entailed certain consequences which could be justly considered paradoxical. This transformation implies, in effect, that an absolute time does not exist on one hand, that two observers in relative motion do not keep the same time, and that on the other hand the spatial separation of two material points no longer has an absolute character and is not the same for the two observers in question. By assuming, as postulates, the absolute character of time and of a distance in space, we necessarily arrive at the Galilean transformation; the adoption of the Lorentz transformation implies the abandonment of these postulates which had seemed so natural. In order to clear up this difficult question, Einstein has made a critical analysis, complete with methods which permit an experimental determination of durations and distances. In order to make this analysis, he had assumed as a fundamental hypothesis that no transfer of energy, no signal, can occur with a velocity greater than that of light *in vacuo* and that, besides, the velocity of light *in vacuo* is a constant independent of the direction of propagation. He was then able to show that the existence of an upper limit of the velocity of propagation for the most rapid signals permitted one to understand and to justify the validity of the formulae of the Lorentz transformation.

First of all, Einstein wondered how, in a system of reference, one should establish a synchronization between the various clocks which measure time at the various points of this system. Since it is impossible to compare directly clocks which are not at the same place, the synchronization of the clocks must be made by an exchange of signals and it is easy to state exactly how this must be done. After a synchronization has been made between all the clocks in the system under consideration, it can be said that henceforth the system has its "proper time." But the synchronization thus obtained is valid only for the system in which it has been made (or for systems at rest with respect to it); it is not allowed, which is the absolutely new result of the Einstein analysis, to define an absolute time valid in all systems. Let us make this point precise. Let there be two systems of reference A and B in rectilinear and uniform motion with respect to each other. We will suppose that the synchronizations of clocks has been made in each of the two systems. The different points of system A are then provided with clocks synchronized with each other, and similarly, the different points of system B are provided with clocks synchronized with each other. As a result of the motion, the clocks of A and the clocks of B pass by each other. If some observers are placed in system A near the clocks and if these observers mark the time shown by the clock in B which is passing before them when their own clock shows a certain time, let us say noon, it will be found that the times thus observed by the different observers on the moving clocks are different. In other words, the times thus observed on the different clocks of B at the same instant of the proper time of A are different. And in the same way, for everything is reciprocal between A and B, the times observed on the clocks of A at the same instant of time of B, by the observers connected with B, are different. There is no longer a simultaneity in the theory of relativity having an absolute sense and valid for systems in relative motion. And Einstein has shown very well how this fact, at first paradoxical, is a consequence of the impossibility of effecting synchronizations with the aid of signals having a velocity greater than that of light *in vacuo*.

Continuing in this way the physical interpretation of the Lorentz transformation, Einstein has shown that everything takes place as if a material object in motion with respect to an observer would seem to him to be shorter in the direction of motion than it would seem to an observer carried along with its motion. In other words, let us consider two observers who are endowed with respect to each other with a rectilinear and uniform relative motion in a certain direction D and let us suppose that one of these observers carries with him a rule oriented parallel to D and whose length, as measured by him, is one meter for example; the other observer will attribute to this rule a length less than a meter, the difference being that much more noticeable as the relative motion is more rapid. Moreover, this "contraction" of the rule relative to the second observer has in general an extremely small value, and it is only when the relative velocity becomes near that of light *in vacuo* that it can be noted. That is the reason why one can not obtain evidence of it directly by experiment, but this contraction, in practice always small, is precisely equal to the contraction imagined by Fitzgerald and Lorentz and is sufficient to explain rigorously the negative result of the Michelson experiment. There is nevertheless an essential difference between the Fitzgerald-Lorentz contraction and that which results, according to Einstein, from the Lorentz transformation. The former, in effect, was considered to be a real contraction brought on by the absolute motion of a body in the ether, while the latter is an apparent contraction relative to the second observer: it is derived uniquely from the manner in which the different observers make their measurement of distances and durations and from the Lorentz transformation which expresses mathematically the relation between the measurements thus made by the two observers.

The apparent contraction of lengths has as a complement an apparent slowing of the clocks. The observers connected with system A, studying the motion of a clock belonging to system B, find that this clock runs slow with respect to their own clocks; they have the impression that the moving clock is losing time. Einstein has shown that this too is a consequence of the Lorentz transformation. The contraction of lengths and the slowing down of clocks are appearances which result from new definitions of space and time with which the Lorentz transformation is connected. And inversely, if the contraction of lengths and the slowing of clocks is postulated, the formulae of the Lorentz transformation are verified.

The arguments by which Einstein has justified his new conception of space and time are often subtle and intricate to develop correctly. But they are perfectly solid and from a logical point of view no serious criticism can be directed against them. In particular we can completely justify the paradoxical fact that the contraction of rules and the slowing down of clocks are reciprocal appearances, i.e., if two observers in uniform relative motion are each supplied with a rule and a clock, the two rules and the two clocks being of identical construction, each of the observers finds that the rule of the other is shorter than his and that the clock of the other runs slower than his own. However surprising this reciprocity must seem at first, it is easily explained when the theory is examined with care, a thing which we naturally can not do here.

The modification of the ideas of space and time, brought about by the Einsteinian principle of relativity, leads to a modification of the principles of kinematics. In particular, there results from it a law of the composition of velocities that is a little more complicated than the classical law. It was one of the successes of the theory of relativity to be able to interpret immediately by means of this new law for the composition of velocities, the result of the experiment of Fizeau on the propagation of light in moving dispersive media. The result of the experiments could be explained in the language of the ether theory by saying that there was a partial dragging of the ether by the movement of the refracting body. Fizeau verified a formula proposed by Fresnel which gave this partial dragging as a function of the refractive index of the moving body. Lorentz with his electron theory had succeeded in justifying this formula, but the theory of relativity has given it an explanation, which is a great deal simpler and more elegant, by showing that it results immediately from the new formula for the composition of velocities.

2. *Space-Time*

THE Galilean transformation was founded on the hypothesis of a complete independence of time and space from which the absolute character attributed to these two notions results. On the contrary, the very form of the equations of the Lorentz transformation shows that in the theory of relativity it is no longer permissible to consider the coordinates of space independently of the coordinate of time. In order to represent geometrically the relations between the coordinates of space and of time for different observers, it is therefore necessary to consider a continuum of four dimensions which in an abstract way realizes the intimate union of space and time, implicitly contained in the Lorentz transformation. It is this geometrical representation which has been developed by Minkowski and which is known under the name of space-time.

The Lorentz transformation leaves invariant the distance between two points of space-time, and the theory of relativity leads to writing all the laws of physics in the form of relations between tensors of space-time. Each observer cuts out, in a certain fashion, his space and his time from the space-time continuum of four dimensions and the formulae of the Lorentz transformation result immediately from the different way in which two observers in uniform motion make this separation of space and time.

Thus the theory of relativity leads to combining, to fusing somehow, into a single continuum the three coordinates of space and the coordinate of time whose physical aspect is so different. We must not, of course, conclude from this that the theory of relativity has thereby come to an identification of time and space. Not only do space and time remain essentially heterogeneous by their very physical nature, but this heterogeneity is treated very clearly in the mathematical theory of Minkowski space-time by the fact that time does not play the same role as the coordinates of space. If we want space-time to be an Euclidean space in the sense of geometry, it is necessary, in order to set up this continuum, to join to the three coordinates of space not the coordinate of time itself, but this coordinate multiplied by $\sqrt{-1}$. That is the mark of the fundamental heterogeneity of space and time.

Besides, time possesses the fundamental property of always flowing in the same sense. From this there results a sort of polarity of space-time, the positive sense of the axis, along which time can be measured, being privileged. A material point i represented at each instant by a point in space-time, and th series of its successive positions in the course of time defines line in space-time, the world-line of the material point. On eacl world-line there is a privileged sense, the sense which goes fron the past to the future, and this unique sense, in which world lines are traced out, succeeds in showing at what point tim differs from space.

But it is no less true that time and space, no matter how dif ferent they are, are no longer independent in the relativity the ory, and four dimensional space-time symbolizes their interde pendence, and constitutes the new framework in which th natural laws must be expressed.

We do not wish to dwell any longer on space-time, a mor detailed study of which could not be made without the aid o mathematical symbolism, and we now wish to show why an how the theory of Einstein has brought about the modificatio of the laws of mechanics.

3. Relativistic Dynamics

THE CLASSICAL EQUATIONS of Newtonian mechanics are invari ant for the Galilean transformation. As long as this transforma tion had been considered as giving the relation existing betwee the coordinates of two observers in uniform relative motion, i followed from this that the equations of Newton were valid i all systems of reference in uniform, rectilinear motion with re spect to the fixed stars. For all observers connected with one o these systems, the mechanical phenomena followed exactly th same laws and no mechanical observation within the system wa able to permit a determination of its absolute motion. That then, was the principle of relativity of the old mechanics. But when Einstein substituted the Lorentz transformation for tha of Galileo to represent the change of coordinates betwee systems in relative uniform motion, the situation was changed

Owing to this substitution, the principle of relativity became valid for optical and electromagnetical phenomena in agreement with the negative result of the Michelson experiment and analogous experiments. But, the equations of Newtonian mechanics were not invariant for the Lorentz transformation; the principle of relativity was no longer applicable to mechanical phenomena, at least not rigorously. Einstein rightly considered this conclusion as inadmissible and assumed that the principle of relativity should be valid for all physical phenomena. But then it was necessary to modify the equations of mechanics so as to make them invariant for the Lorentz transformation, this modification, of course, being made in such a way that Newton's equations remain valid as first approximations in all the usual cases where their application has led to such brilliant results. It was easy to find the new form to give these fundamental equations of mechanics to make them invariant for the Lorentz transformation. Newton's equations say that the derivative of the momentum, with respect to time, is equal to the force. The dynamics of Einstein keeps this statement, but gives momentum a definition different from that which classical dynamics adopted. In place of setting the momentum of a material point equal to the product of the mass by the velocity, the new dynamics sets it equal to the product of the mass by the velocity, divided by a factor which is a function of the velocity. As long as the velocity is small enough so that the ratio of its square to the square of the velocity of light *in vacuo* is negligible compared with unity, this factor can be taken equal to one without sensible error and we then have the old formula again. But for velocities that are large, comparable to that of light *in vacuo*, the factor in question is no longer equal to one, and varies with the velocity. There then results differences with respect to the classical laws, differences which become more and more sensible as the velocity approaches that of light. Moreover, it is easy to deduce from the new equations of dynamics that a material point can never attain the velocity of light *in vacuo*. The velocity of light *in vacuo* thus seems as the upper limit of the velocity of transport of energy in space. So we find justified *a posteriori* one of the hypotheses introduced by Einstein in his critique of the method of synchronizing clocks.

We can not enter here into the details of the equations of relativistic mechanics. Let it suffice us to say that this mechanics can be developed following exactly the same scheme that had succeeded so well in classical mechanics. For example, all the new dynamics can be deduced by starting out from a principle of stationary action from which the equations of Hamilton and Lagrange were deduced: the principle of Least Action of Maupertuis in the case of constant fields, and the theory of Jacobi are again found. But there is one essential difference between the old and new mechanics: the function, which figures in the integral of action, is not the same. Nevertheless, the relativistic function of action approaches the classical function of action whenever the velocity of the moving object is small enough so that the square of the ratio of this velocity to the velocity of light *in vacuo* is negligible. An immediate consequence of this is that classical mechanics is a perfectly valid approximation in the majority of usual cases.

The modification introduced into the equations of mechanics by relativistic conceptions can, as we have seen, be expressed by the fact that the momentum of the material point is the product of a constant characteristic of the material point by its velocity, divided by a certain function of the velocity. One can, if one wishes, say that the momentum of a material point always is, as in classical mechanics, the product of the mass by the velocity, but on condition that the mass is considered as varying with the velocity. Since the function appearing in the denominator in the expression for the momentum tends toward unity as the velocity tends toward zero, the constant of the numerator is the mass of the material point at rest: it is often called the "proper mass" for it is the mass of the point for an observer associated with it. As we have already said, the variation of mass with velocity becomes sensible only for velocities near that of light *in vacuo*.

The modification brought by relativity to the expression for momentum is accompanied by a corresponding modification in the expression for energy. This is not surprising for it can easily be shown that the three components of the momentum and the energy constitute the four components of a four-dimensional vector in space-time, the world force 4-vector. Momentum and energy thus constituting a single mathematical idea, it is not surprising that a modification of one should be reflected in the other. The new expression obtained for the energy has this interesting property: it does not become zero when the velocity becomes zero, but has a constant value equal to the product of the proper mass by the square of the velocity of light *in vacuo;* any material point, any body endowed with inertia, thus possesses a proper energy independent of its velocity. If the velocity is not zero, the energy of the body is greater than its proper energy, and the difference between the total energy of the body in motion and its proper energy is the energy due to the motion, which can be called the kinetic energy. If the relativistic expression for the kinetic energy is examined, we see that for velocities small with respect to that of light, it is sensibly the value foreseen by classical mechanics (one-half the product of the mass by the square of the velocity). Here then we again find the character of a first approximation valid for velocities small, compared to that of light, which, in the eyes of the relativists, justifies the usual use of the Newtonian formulae.

A material body at rest with respect to an observer possesses for this observer an energy equal to the product of its proper mass by the square of the velocity of light. But we have seen that if the body is in motion, its mass has a value, depending on its velocity, which is indistinguishable from its proper mass for small velocities and moreover tends to infinity when the velocity tends to that of light. It can be shown that the energy of a body evaluated by an observer is always equal to the product of the square of the velocity of light by the mass of th body in motion. The energy of the moving body thus tends tc infinity when its velocity tends to that of light, a new aspec of the impossibility of communicating to a body a velocity equal to or superior to that of light *in vacuo*. Einstein ha generalized this result by showing that all bodies, all physica entities, possessing a certain mass (evaluated by a certain ob server) possess, by this very fact, a certain energy which, evalu ated by the same observer, is equal to the product of this mas by the square of the velocity of light. He has illustrated this statement by numerous examples. Thus there is established by this "principle of the inertia of energy" an intimate and genera correlation between mass and energy. From this it follows, that all bodies lose mass if they lose energy, and inversely, they ac- quire mass if they absorb energy. Thus, for example, an atom emitting radiation finds its mass diminished. The principle of the inertia of energy has, since its announcement, played an impor- tant role in the questions of theoretical physics as much so in astrophysics as in nuclear and atomic physics. In particular, it has contributed powerfully to the writing of the energy balances in the phenomena of disintegration and in establishing the for- mulae of the reaction between nuclei which represent these phenomena. But these are questions which we must now lay aside.

4. General Relativity

WE SHALL SAY here only a few words concerning the theory of General Relativity. At the beginning of the development of his theory, Einstein was preoccupied only with systems of coordi- nates which are in rectilinear and uniform motion with respect to the fixed stars. He thus obtained a form of the principle of relativity applicable to rectilinear and uniform motion as was that of classical mechanics. Hence, the name of Special Relativ- ity was given to the body of results that he first announced, and concerning which we have recalled the most essential points. But it was evidently necessary to try to generalize these results and to establish a theory applicable to motions which are not rectilinear and uniform. For these motions there is in general no longer a principle of relativity in the special sense of the word, for to an observer bound to an accelerated system, to a system in rotation for example, there is an influence of the motion on the course of the mechanical, optical and electromagnetic phenomena. In particular, the mechanical phenomena in the accelerated system can be calculated only by bringing in fictitious forces called "centrifugal forces" and "Coriolis forces" and the effects produced by these forces reveal, to the accelerated observer, that he is not at rest. Nevertheless, the idea of relativity can be kept under the most general form, by assuming that the laws of nature are always expressed by tensorial relations in space-time, and by trying to explain the effects of acceleration on the physical phenomena by the very fashion in which the coordinates of the accelerated observer are defined. By making this analysis, we see that the accelerated observer employs curvilinear coordinates in space-time and this is enough to explain the appearances which he observes, notably the appearance of centrifugal forces and centrifugal components.

While investigating this problem thoroughly, Einstein had a splendid idea which led him to his celebrated theory of gravitation. The force of gravity, which plays so important a role in our interpretation of astronomic facts, has always stood a little apart from the other natural forces with which we are familiar. It has this essential character of always being proportional to the mass of the attracted body, and the very precise experiments of Eötvös have shown that this proportionality is quite exact. Hence, it follows from the very form of the equations of dynamics, that the motion of a body in a purely gravitational field is independent of its mass. Trajectories are then determined in it without our needing to know the nature of the body destined to follow these trajectories. These trajectories result somehow from the intrinsic properties of the field of gravity. Einstein has seen in this fact proof that the existence of a gravitational field in a region of space implies the existence of a local curvature of space-time. The space-time of Special Relativity is, indeed, a continuum of four dimensions included in the category of Euclidean continua of which the plane is an example in two dimensions. But nothing prevents us from supposing that space-time is not everywhere Euclidean, that it possesses local curvatures. There does not then exist in this space, systems of rectilinear Cartesian coordinates, and the position of points in it can be determined only with the aid of coordinates analogous to those that are used in geometry for the study of curved surfaces. Observers placed in curved regions of space-time therefore necessarily use curvilinear coordinates to determine events in it, and from this follows the appearance of gravitational forces. While the centrifugal forces present in a rotating system are due to the fact that an observer bound to this system uses a system of curvilinear coordinates in order to mark out events in Euclidean space-time, gravitational forces are due to the fact that, where a gravitational field is present, space-time is curved, and that an observer is obliged to use curvilinear coordinates in it. I shall be satisfied here with this rapid indication of the theory of gravitation of Einstein which can be further developed only with the aid of a complicated mathematical apparatus. Let us say only that it is a quite coherent theory and completely satisfying to the intellect.

Special Relativity has received good experimental confirmations, for the variation of mass with velocity, predicted by the dynamics of Einstein, and which should be significant for electrons endowed with velocities comparable to that of light, has been effectively verified by a series of experimental researches, of which the most recent and most decisive are those of Guye and Lavanchy. Likewise, the principle of the inertia of energy has been of too great service, notably in nuclear physics, for one to doubt its validity. But if Special Relativity seems well confirmed by experiment, it is proper, we believe, to be a little less affirmative concerning General Relativity. The new phenomena predicted by it are indeed very small and, even when they are actually observed, it can always be asked if they really have their origin in the cause which the theory of Einstein attributes to them, or rather in some other very small perturbation which was neglected in the analysis. Neither the very small secular displacement of the perihelion of Mercury, nor the very small deviation of light rays passing near the solar disc, seems to bring irrefutable proof of the correctness of the relativistic conceptions concerning gravitation: these phenomena exist and have the order of magnitude predicted by the theory of Einstein, but their interpretation is not set in any absolute fashion. More convincing seems to be the displacement toward the red of the spectral lines emitted by the companion of Sirius, but a single verification of this kind can not be considered sufficient.

Whatever the experimental verifications of the General Theory of Relativity may be, the body of the theory of Einstein forms an imposing monument. This theory brought us a host of new and fruitful ideas. It has accustomed us to reject preconceived ideas, to scrutinize with care the very bases of our theoretical conceptions. By its very difficulty, the study of the theory of relativity has been a wonderful exercise in adaptation for the minds of the theoretical physicists.

Chapter 5 | The Appearance of Quanta in Physics

1. Classical Physics and Quantum Physics

THE MOMENT has come for us to speak of the appearance of quanta in physics, but before recounting the story of this appearance, it will be useful to show, in a few words, the profound differences which contrast classical, prequantum physics, of which we have spoken in the preceding chapters, and the quantum theories which now are going to hold our attention. The common postulate behind all the theories of classical physics is that it is possible to represent the state of the physical universe by elements distributed in the framework of three dimensional space and evolving in a continuous manner in the course of time. The motion of the physical elements is defined by the fashion in which their position varies in the course of time. Assuredly there is an essential difference between the relativisitic conception and the preceding conceptions. In prerelativistic physics, space forms a fixed framework in which the physical phenomena observed by all imaginable observers are going to take place, a single universal and absolute time impressing its rhythm on all these observers. For the relativist, on the contrary, neither space nor time has an absolute character: this character is possessed only by the four-dimensional continuum formed by the fusion of space and time and called space-time. In the space-time continuum, the various observers cut out their space and their time in different fashions. Despite this essential modification of the concepts of space and of time, the relativist is in accord with his predecessors in asserting that each observer can represent the totality of physical phenomena in a framework of space and of time which is well defined and completely independent of the nature of the entities which enter into it. Thus, a given observer will always be able to represent the existence of a corpuscle by a well-defined sequence of positions in space, successively occupied in the course of time, without having to occupy himself with the physical nature of this corpuscle, with the value of its mass for example. Moreover, for the relativist, as for the physicist of the preceding epoch, the entire evolution of the phenomena is governed by an inexorable play of differential equations which determine the entire future; in subscribing to space-time, the relativist presumes given the entire totality of events corresponding to the whole course of time, and for him, it is because of an insufficiency of the human intellect that an observer can discover the events contained in space-time only by successive sections, and in proportion to the flow of his own proper time. By his affirmation of the possibility for each observer to localize events exactly in space-time, by the spatialization of duration, and the negation of all real becoming, that the very conception of space-time implies, the theory of relativity, while pushing them to their extreme consequences, retains the guiding ideas of the old physics. Thus, it can be said that, despite the so-new and almost revolutionary character of the Einsteinian conceptions, the theory of relativity is in some ways the culmination of classical physics.

But the orientation of present quantum theories is quite different. We have already indicated in the introduction of this work several of the essential traits of the quantum theories. The existence of the quantum of action, as we have said, implies a sort of interdependence between the localization of an object in space and in time, and its dynamic state. This was completely unsuspected by classical physics and is even more astonishing in its consequences than the connection established by the theory of relativity between the variables of space and the variable of time. This interdependence has, as a consequence, the impossibility of simultaneously determining position and motion, an impossibility for which the Uncertainty Relations of Heisenberg are a precise expression; it implies the impossibility of making experiments and measurements which permit us to make the localization in space-time and the dynamic state simultaneously precise. In scrutinizing this critical question, we see that the framework of space and of time used by the old physics (or even the framework of space-time of relativity physics) from the quantum point of view is only an approximation valid for heavy bodies. And by heavy bodies, we here understand bodies composed of an enormous number of elementary particles and consequently endowed with a total mass, very large with respect to that of the particles: in this category of heavy bodies we immediately place all the bodies which we perceive directly in our usual experience. This explains why classical physics directed to the study of phenomena on our scale could be content with this framework of space and time. Axes of reference drawn on a material body, a clock set up in the usual fashion, permit us to define coordinates of space and of time capable of being used for a practically perfect description of macroscopic phenomena, in accordance with the conceptions adopted by classical physics. But if it is desirable to describe the evolution of the microscopic world, the history of elementary corpuscles, with the aid of the coordinates of space and time so defined, we run headlong into the uncertainties of Heisenberg, and the existence of these uncertainties warns us that the space and time of classical physics, well-defined and perfectly usable on the macroscopic level, cease to be perfectly adequate for the description of physical reality on the level of atoms and electrons. But those of us who are macroscopic physicists necessarily seek to describe the world of elementary corpuscles in the framework of space and time, which is suggested to us by all our current experience: hence the difficulties we meet in the quantum theory, hence the so-mysterious character that the very notion of the quantum of action presents to us. Perhaps it would be possible to try to construct for the corpuscular world a more general and less precise framework of space and time than the classical one. To be satisfactory, these new notions, which should contain within themselves the quantum of action, and consequently separate the geometrical and dynamic aspects in a less complete manner than the old ones, should allow us to recover for assemblages of an enormous number of corpuscles, in other words for material bodies, our usual, prior conceptions of space and of time. Interesting preliminary work has already been made in this direction by Jean Louis Destouches, and this is a path that must not be lost from view.

The manner in which classical physics conceived the absolute determinism of physical phenomena rested essentially on the manner of thinking about space and time and the theory of relativity, while revising so profoundly our ideas relative to space and to time, had, however, sufficiently respected them so as not to breach classical determinism. It is not the same with the quantum theory. The latter, by not allowing us any longer to represent the evolution of individual phenomena in a continuous manner in the framework of space and time, obliges us to abandon determinism completely, or at least to modify profoundly the conception we had of it. The impossibility of knowing at the same time the configuration and the dynamic state of the elements of the microscopic world, an impossibility flowing from the existence of the quantum of action, acts so that the several successive observations of the microscopic world that we are able to make never let us know all the elements we have need of to establish between the results of these observations a rigorous bond consistent with the scheme of classical determinism. Indeed, the present quantum theory furnishes us only with laws of probability, allowing us to say what, the result of the first observation being given, is the probability that a later observation will furnish us such and such a result. This substitution of laws of probability for rigorous laws in the microscopic domain is certainly bound up with the non-validity of the concepts of space and time in this domain. For objects in the macroscopic domain, the notions of space and time take on again their validity in a certain asymptotic way and it is the same for determinism, the probability of the predictions permitted by the quantum laws tending there to certainty.

What we have just recounted suffices to show what an essential step theoretical physics had to take that day when it recognized the necessity of taking account of the quantum of action. And now it is necessary for us to show how this event took place thirty-five years ago.

2. *The Theory of Black Body Radiation and the Quantum of Planck*

THE ORIGIN of the quantum theory lies in the researches made around 1900 by Max Planck on the theory of black body radiation. This theory, developed with the help of the methods then in use in physics, had led to grave difficulties. This is what we must first explain.

If we consider an enclosure maintained at a uniform temperature, material bodies contained in this enclosure emit and absorb radiation, and finally there is established a state of equilibrium in which the exchanges of energy between matter and radiation balance each other. Relying solely on the fundamental principles of thermodynamics, Kirchhoff had shown that this state of equilibrium is unique and corresponds to a perfectly well-determined spectral distribution of the radiation enclosed in the enclosure. Moreover, the distribution of this radiation depends on the temperature of the enclosure only, and in no way on the dimensions or form of the enclosure or the particular properties of the material bodies that it contains or the composition of the walls. This equilibrium radiation, characteristic of the given temperature, is often called by the incorrect name of the "black body radiation" corresponding to this temperature.

Therefore for theoretical physics it was an essential task to predict the spectral distribution of the black body radiation corresponding to a given temperature. The problem was at first attacked with the aid of methods which relied principally on the principles of thermodynamics and thereby possessed a character of very great sureness. It could first be shown that the total density of the black body radiation (that is, the quantity of radiant energy contained in a unit volume in the interior of the enclosure in thermal equilibrium) is proportional to the fourth power of the temperature measured on the absolute scale. This is the Stefan-Boltzmann Law. Then by more extended reasoning, Wien showed that the spectral density of the black body radiation for a certain frequency ought to be proportional to the product of the cube of the frequency by a function of the quotient of the frequency by the temperature. Unfortunately, this last function is not determined by the thermodynamic reasoning of Wien. The laws of Stefan and of Wien give important information on the composition of the radiation and on its modifications as a function of the temperature, and experiments have verified them thoroughly, but they do not completely fix the form of the law of spectral distribution. It finally seemed that one could not go any further by appealing only to thermodynamic considerations and that, in order to determine completely the form of the law of spectral distribution, it would be necessary to introduce hypotheses on the manner in which matter emits and absorbs radiation, that it would consequently be necessary to venture onto the field of atomic hypotheses while abandoning the more solid terrain of thermodynamics.

However, this did not seem to be difficult, for the electromagnetic theory, particularly in the electron form of Lorentz, offered a seemingly satisfactory image of the mechanism of emission and absorption of radiation by matter. One had only to use the formulae that it furnished in order to be able to obtain the form of the function left indeterminate by the reasoning of Wien, and thus make the spectral distribution of black body radiation completely precise. The results of this work were quite disappointing. The law of spectral distribution thus obtained (Rayleigh's law) was not in accord with experiment. It predicted a monotōnic increase of the spectral density with the frequency while experiments clearly indicated that the spectral density, after attaining a maximum for a certain frequency thereafter diminished indefinitely as the frequency increased, a fact that can be interpreted by saying that the curve representing the variations of the spectral densities is a "bell-shaped" curve. Because it predicted an indefinite increase of the spectral density with frequency, Rayleigh's law resulted in an absurd conclusion: the total density of the black body radiation at any temperature should be infinite!

The situation created by this divergence between the theoretical predictions and the facts was particularly grave for, the more the physicists worked over the demonstrations of Rayleigh's law, the more they saw that this law is an inescapable consequence of the body of classical theories. It can be shown (Jeans) that Rayleigh's law can be obtained by taking into account the number of stationary waves capable of existing in the enclosure and by applying to them the general results of classical statistical mechanics. All hope of arriving at a law of black body radiation other than that of Rayleigh and consistent with experiment thus disappeared unless quite new points of view were introduced into natural philosophy. It is to Max Planck that the honor of realizing this revolution came.

Planck began a re-examination of the question by imagining that matter is formed of electronic oscillators, that is, of electrons capable of oscillation around a position of equilibrium under the action of a force proportional to the displacement. He studied the equilibrium which is realized in an isothermal enclosure as a consequence of the exchanges of energy between the oscillators and the ambient radiation. Since the composition of the equilibrium radiation should be independent of the nature of the material bodies present in the enclosure, the result obtained by this method ought to have a general validity. By analyzing the exchanges of energy between the radiation and the oscillators according to classical methods, Planck naturally found again the law of Rayleigh. But in examining this result, he was able to see that the inexactness of this law comes from the too great role which, in the classical picture of the exchanges of energy between oscillators and radiation, the oscillators of high frequency play. It is indeed the importance of the exchanges of energy between the equilibrium radiation and the material oscillators of high frequency that leads to a monotonic increase of the spectral density with frequency and to the experimentally inexact or logically absurd consequences that we have noted above. Planck then had the brilliant idea that it was necessary to introduce into the theory a new element, entirely strange to classical conceptions, which would act to restrain the role of the oscillators of high frequency, and he set down the following famous postulate: "matter can emit radiant energy only in finite quantities proportional to the frequency." The proportionality factor is a universal constant, having the dimensions of mechanical action. It is the celebrated constant h of Planck.

Bringing into play this hypothesis of paradoxical aspect, Planck took up again the theory of thermal equilibrium and found a new law of spectral distribution of black body radiation to which his name has become attached. Since nothing in the premises of Planck is contrary to the principles of thermodynamics, the formula that he obtained is in accord with the laws of Stefan and Wien. On the other hand, it coincides with the formula of Rayleigh only for low frequencies and high temperatures, but it departs from it completely for high frequencies and low temperatures. This is easily understood. For low frequencies and high temperatures, the energy exchanges between matter and radiation bring into play a very great number of very small "grains" of energy: everything then takes place sensibly as if these exchanges are made in a continuous manner and the classical arguments should lead to results which are sensibly exact. On the other hand, for high frequencies and low temperatures, the energy exchanges bring into play a small number of large grains of energy, and the classical arguments should not apply. Hence Planck's law of spectral distribution departs completely from that of Rayleigh for high frequencies and low temperatures. In particular, while, for a given temperature of the enclosure in thermal equilibrium, the law of Rayleigh foresees a monotonic and inadmissible increase of the spectral density with frequency, Planck's law foresees a density which, after having increased with the frequency, passes a maximum and thereafter diminishes indefinitely for very high frequencies. The curve which represents the variation of density as a function of the frequency according to Planck's law is a bell curve. Hence it is easy to see that the total density of the black body radiation has a finite value, and thus one of the grave difficulties of the classical theory is found to be eliminated.

The numerical comparison of the new law of the spectral distribution of black body radiation with the experimental results, whose number and precision was then increasing without let-up since the attention of physicists was attracted to this question, permitted Planck to show that the facts are entirely in accord with the formula furnished by his theory, on the condition of attributing to the new constant h a perfectly well-determined numerical value. This numerical value calculated by Planck is extremely small when it is expressed in the usual physical units. It is quite remarkable that the numerical value of the constant h should have thus been obtained at the first attempt, with a great exactness, with the help of only those data relative to black body radiation. Since then the appearance of the constant h has been found in a considerable number of physical phenomena of a very different nature and thus a great number of independent methods of measuring this constant have been found. These different measurements, of increasingly greater precision, have always given values very close to that which was found at the beginning by Planck by means of a single phenomenon.

It is probable, that at the time when Planck wrote his fundamental papers on the theory of black body radiation, contemporary physicists did not at once comprehend the importance of the revolution which had just been accomplished. The hypothesis of Planck without doubt must have seemed to them an ingenious means of improving the theory of an interesting but, all in all, a particular phenomenon, and not a brilliant idea destined to upset all the classical conceptions of physics. But, little by little, the fundamental importance of the idea of Planck became apparent. The theorists perceived that the discontinuity introduced by the hypothesis of quanta is incompatible with the general ideas which served up to then as the bases of physics and demanded a complete revision of these ideas. We can not admire too much the intuition of genius which permitted Planck, by the study of a particular physical fact, to perceive at a glance one of the most fundamental and mysterious laws of nature. More than forty years have passed since this marvelous discovery and we are still far from having achieved a comprehension of all its import, and exhausting all its consequences. In the history of the progress of the human mind, the conquest of the constant of Planck must remain a memorable date.

3. Development of the Hypothesis of Planck. The Quantum of Action

PLANCK had reasoned, in his theory of radiation in thermal equilibrium, by supposing that matter contains electronic oscillators through the intermediary of which the exchanges of energy are made between matter and the ambient radiation. Now, an oscillator, that is, a material point drawn back to its equilibrium position by a force proportional to the displacement, constitutes a mechanical system displaying a very special property: its oscillations have a frequency independent of the amplitude. In other terms, the oscillator has a unique frequency which is always the same, no matter what the intensity of its motion of vibration. For each oscillator there can then be defined, according to Planck, a "quantum of energy" equal to the product of the frequency of this oscillator by the constant h and this has a definitely determined meaning which says that when there are exchanges of energy between the oscillators and radiation, the oscillators can only gain or lose a finite quantity of energy equal to their quantum. But this hypothesis of the quantum of energy has the inconvenience of being valid only for harmonic oscillators. If any kind of mechanical system capable of vibrating periodically is considered, the frequency of the motion of vibration depends in general, on the intensity of the motion, and the system does not have a well-defined quantum of energy. Planck felt the necessity of stating the hypothesis of quanta in a general form which would be applicable to all mechanical systems and which in the particular case of an oscillator would again give the statement of the quantum of energy. In order to arrive at such a statement, he noted that the constant h having the dimensions of an action (i.e., of energy multiplied by time or momentum multiplied by length) plays the role of an elementary quantity of action, a kind of atom of action. Now if there is a periodic motion that can be defined by means of a single variable, for example, the periodic motion of a corpuscle along a line, we can calculate the integral of the Maupertuisian action, the same which appears in the statement of the principle of Least Action, for a complete period of the motion. It is a characteristic constant of the periodic motion and, by putting this constant equal to a whole multiple of the constant h, we obain a new statement of the hypothesis of quanta which has the advantage of being applicable to any periodic motion defined with the aid of a single variable. Moreover, we can easily verify that this new statement in the particular case of the linear oscillator leads to the original statement of Planck. It might be said that in order to find a general form of his theory, Planck had to give up the original hypothesis of the quantum of energy and substitute for it the hypothesis of the quantum of action.

This appearance of action in the correct statement of the quantum hypothesis was at once logical and surprising. It was logical because classical mechanics indicated in its principles of Hamilton and of Least Action the important role of this quantity, and because the theories of analytical mechanics where action appears, offer a sort of framework already prepared for the introduction of quantization. But, on the other hand, it was also very surprising, for it is very difficult to understand from the point of view of physics how a quantity like action, whose abstract character is very prominent, and which obeys no theorem of conservation, can present a sort of atomicity. Action is always expressed by the product of certain quantities of a geometrical character and certain quantities of a dynamic character which correspond in pairs and form the canonically conjugate variables of analytical mechanics. Thus the integral of Least Action of Maupertuis is the line integral of the momentum along the trajectory. The kind of atomicity of action expressed by the appearance of the constant h implies then the existence of an interdependence between the framework of space and time and the dynamic phenomena which we try to localize in it. This interdependence has a quite new character and is quite foreign to the conceptions of classical physics. Hence the profoundly revolutionary character of the hypothesis that Planck, in a stroke of genius, had put at the base of his theory of black body radiation.

Planck had posited in principle that matter can emit radiation only by finite quantities, by grains. This does not necessarily imply a discontinuous structure of radiation once emitted, for it is possible to develop the theory in two different manners which lead to opposite conceptions on the subject of the absorption of radiation by matter. The first attitude, one would say the most frank, and which triumphed in the end, consists in supposing that the elements of matter, the electronic oscillators, for example, can take only certain states of motion corresponding to the quantized values of the energy, from which it follows that for absorption as well as for emission, the exchanges of energy between matter and radiation can be made only through quanta. But then it necessarily follows that radiation has a discontinuous structure. Recoiling from this formidable consequence of his own ideas, Planck for a long time made the greatest efforts to develop the second less radical form of the quantum theory in which only the emission would be discontinuous, the absorption remaining continuous. Matter would be capable of accumulating in a continuous manner a part of the radiant energy which would fall on it, but it would not be able to emit it except by spurts and in finite quantities. We can easily understand the goal of Planck's efforts: he wished to safeguard the continuous nature of radiation because it alone seemed reconcilable with the wave theory which rested on innumerable verifications of extreme precision. Despite all the ingeniousness brought by Planck to the development of this form of the quantum theory, it found itself undermined by the further progress of physics, in particular by the interpretation of the photoelectric effect and by the success of the theory of the atom of Bohr. Turning now to the first of these questions, we are going to see how Einstein, by interpreting the photoelectric effect in conformity with the spirit of the quantum theory, found himself led back to a corpuscular theory of light.

4. *The Photoelectric Effect and the Discontinuous Structure of Light*

THE DISCOVERY and the study of the photoelectric phenomenon held a very great surprise for physicists. This phenomenon consists in this: that a piece of matter, exposed to the action of radiation with a sufficiently short wave length, often expels electrons in rapid motion. The essential characteristic of this phenomenon is that the energy of the expelled electrons is solely a function of the frequency of the incident radiation and does not depend in any way on its intensity. Only the number of the electrons depends on the incident intensity. These simple empiric laws made most difficult the theoretical interpretation of the elementary mechanism at the base of the liberation of the photoelectric electrons, or photoelectrons, as they are called today. The wave theory of light, which in 1900 seemed to rest on inescapable bases, leads to the consideration of the radiant energy as being distributed uniformly in the light wave. An electron struck by a light wave therefore receives the radiant energy in a continuous fashion, and the quantity of energy that it thus receives in a second is proportional to the intensity of the incident wave, and depends in no way on the wave length. Hence the laws of the photoelectric effect seem difficult to explain.

In 1905 Einstein had had the very remarkable idea that the laws of the photoelectric effect indicate the existence of a discontinuous structure for light where quanta would make their appearance. The hypothesis of Planck, in its first form, the most straightforward, consists in assuming that radiant energy can be absorbed by matter only in finite quantities proportional to the frequency: the success of the theory of black body radiation of Planck had shown the solid basis of this hypothesis. But if this hypothesis is correct, it would seem very probable that this granular structure of radiation, if it is manifested at the moment of emission and at the moment of absorption, ought also to exist in the intermediate period when the radiation is being propagated. Einstein had accordingly assumed that all monochromatic radiation is divided into grains whose energy has a value proportional to the frequency, the constant of proportionality being, of course, the constant of Planck. Thereupon the laws of the photoelectric effect are easy to interpret. When an electron contained in matter receives a grain of light, it will be able to absorb the energy of this grain and leave the matter where it is held, on this condition, however, that the energy of the grain of light is greater than the work necessary for the electron to leave the matter. The electron thus expelled by the action of the light will then possess a kinetic energy equal to the energy of the grain of light absorbed, less the work expended in leaving the matter: this kinetic energy will then be a linear function of the frequency of the incident radiation, the slope of the line which represents it as a function of the frequency being numerically equal to the constant of Planck. All these predictions are shown to be in perfect accord with experiment. First, if the frequency of the incident light is varied, it is found that the photoelectric effect is produced only when the frequency is greater than a certain value, the photoelectric threshold. Then, in the frequency band where the effect takes place, the kinetic energy of the photoelectrons is a linear function of the frequency of the incident light, and if the line representing this linear dependence is drawn, the slope is found to be numerically equal to Planck's constant. Naturally in the new granular conception of light, the intensity of the incident light measures the number of grains of energy which fall per second on a square centimeter of surface of the irradiated body, and hence, the number of photoelectric effects which are produced per second in the interior of the body, should be proportional to the intensity.

Such is the interpretation of the laws of the photoelectric effect proposed by Einstein in 1905. He called it the theory of light quanta (*Lichtquanten*). Today we call it the theory of photons, for we have given to the grains of light the name of photons. For thirty years now, the existence of the photon has received numerous confirmations. Not only has the study of the photoelectric effect for light been made with an increasing precision, entirely confirming the relations discovered by Einstein, but the study of this same photoelectric effect when it is produced by X-rays and gamma rays has furnished an even more precise and clearer confirmation of the theory of photons. For X-rays and gamma rays, the frequencies are a great deal higher than in the case of light; the energy transported by each photon is thus a great deal larger and these radiations are capable of detaching, by means of the photoelectric effect, electrons solidly anchored in the depths of the atoms of the irradiated substance. Since the study of X-ray spectra allows us moreover to find out very exactly the work necessary to detach an interior electron from an atom of a known nature, the work necessary to extract a photoelectron can be calculated in this case with a relatively greater precision than in the case of light. The study of the photoelectric effect of X-rays and gamma rays has therefore permitted us to submit the exactness of the photoelectric relation of Einstein to a very rigorous test: the numerical verification has been perfect and the theory of grains of light has been strengthened thereby. (Maurice de Broglie, Ellis, Thibaud . . .)

The discovery of another phenomenon in 1923 has come to furnish a new proof of the existence of the photon. We should like to speak of the Compton effect. We know that if radiation strikes a material body, a part of this energy is, in general, dissipated in all directions in the form of scattered radiation. The electromagnetic theory interpreted this scattering by saying that under the influence of the electric field of the incident wave, the electrons contained in the material body enter into forced vibrations and become the source of small spherical secondary waves which thus scatter in all directions a part of the energy brought in by the primary wave. According to this interpretation, the vibration scattered under the action of a primary monochromatic wave ought to have exactly the same frequency as this primary wave. For a long time, the electromagnetic theory for this scattering was shown to be perfectly adapted to the interpretation of these phenomena, first in the domain of light and then in that of X-rays. The laws predicted by the theory were verified exactly. But a more precise study of the scattering of X-rays by matter showed us that alongside a scattering without change of frequency, as predicted by the electromagnetic theory, there is produced a scattering with a lowering of frequency which is quite impossible to foresee by classical reasoning. It was the American physicist, A. H. Compton, who had the great distinction not only of establishing definitely the existence of this new phenomenon, but of studying its laws in a precise fashion and of proposing an interpretation for it. The essential fact observed by Compton is that the radiation scattered with a lowering of frequency has a frequency which varies with the angle of the scattering, but is independent of the nature of the scattering body. Compton, and at almost the same time, Debye had the idea that these laws could be interpreted by comparing the scattering with change of frequency to a collision between an incident photon and an electron contained in the matter. At the moment of collision there is an exchange of energy and momentum between the photon and the electron, and since the electron can, in general, be considered as almost immobile in comparison with the photon, it is always the photon which loses energy while the electron gains. The frequency of the photon being proportional to its energy, there is a lowering of the frequency at the moment of collision. The theory is developed very simply by relying on the theorems of the conservation of energy and momentum, and permits us to find very exactly the variation of frequency of the scattered photons as a function of the angle of scatter such as was indicated by experiment. The independence of the phenomenon with respect to the nature of the scattering substance, at least in so far as the variation of wave length in concerned, is immediately explained by the fact that the phenomenon brings into play only the properties of electrons, the universal constituents of all material bodies. The Compton-Debye theory has explained the essential characteristics of the Compton effect in so complete and fortunate a fashion that it has given a striking confirmation to the theory of photons.

We can cite as still another confirmation of the conception of photons, the discovery of the Raman effect, a little after that of the Compton effect. In the Raman effect there is a scattering of light with a change of frequency. The phenomenon differs profoundly from the Compton effect in that the change of frequency that the light undergoes at the moment of scattering depends essentially on the nature of the scattering body. Moreover, there generally is some scattering which takes place with an increase in frequency; however, this type of scattering is a great deal less intense than that which is accompanied by a lowering of frequency. The photon theory explains quite well the essential features of the phenomenon, and in particular gives an immediate interpretation of the predominance of the Raman effect, with lowering of frequency, a predominance for which the theories founded on classical ideas could not account.

In brief, for thirty years, the hypothesis according to which light energy should present a granular structure has been shown to be very fruitful and there is no doubt that it reveals to us an essential aspect of physical reality. But it raises difficulties as well, and from the first works of Einstein on this subject, there has been no lack of objections to it. First of all, how are we to reconcile this discontinuity in the structure of light with the wave theory to which so many experiments in physical optics had brought verifications of an extreme precision? How are we to imagine the existence of indivisible grains of light, when interference experiments show that it is possible to obtain continuous wave trains of a length of several meters? It is not possible, as Lorentz has shown, to interpret in a reasonable way the laws of the resolving power of optical instruments, for example telescopes, if it is supposed that light energy is concentrated into localized grains in space. And how are we to understand the very existence of interferences? Surely, it would be possible to conceive of the simultaneous arrival of a great number of grains of light in a certain pattern which could produce the appearance of the phenomena of interference by consequence of the mutual action between the grains. But then the phenomena of interference should depend on the intensity of the light employed, and if this light were weak enough so that on the average there would never be in the interferometer more than one photon, the interferences would disappear. This experiment has been made (first of all by Taylor) and the result of it has been the following: however weak the incident light may be, there is always obtained the same phenomena of interference on the condition, of course, of sufficiently long photographic exposure. This proves that each photon considered by itself must give rise to interferences, which is quite paradoxical if the photon is considered as an isolated and localized grain.

Still other objections show how difficult it is to assume a purely granular conception of radiation. First, the very manner in which Einstein defined the quantum of light brings in a non-corpuscular element: the frequency. A purely corpuscular picture of radiation does not allow us to define a periodicity, a frequency, and, in fact, the frequency which figures in the definition of Einstein, is the frequency of the wave theory, whose value is deduced from phenomena of interference and diffraction. The relation which defines the energy of the photon as being equal to the product of the frequency by Planck's constant can not therefore serve as the basis of a purely corpuscular conception of radiation. It constitutes rather a kind of bridge thrown up between the wave aspect of light, well-known since Fresnel, and the corpuscular aspect revived by the discovery of the photoelectric effect. It would be inexact, though, to say that before the discovery of the photoelectric effect, nothing could have made us think of a corpuscular conception of light. We have already seen at what point rectilinear propagation, reflection at mirrors, and in a general fashion all of geometric optics with its notion of light rays naturally oriented our thought toward an optics of a ballistic character. But the theory of Fresnel, by arriving at a wave interpretation of all these phenomena of a ballistic aspect, had seemed to make the whole granular conception useless. The discovery of the photoelectric effect indicated the necessity of returning to a conception of this kind, but, at the same time, the very form of the Einstein relation showed that it was necessary to unite the granular conception and that of waves in such a manner that the two terms of the relation would have a physical meaning.

It is necessary to point out a still more subtle difficulty. In the classical conceptions, the energy of a corpuscle is a quantity that has a completely determinate value. On the other hand, in the theory of radiation, radiation can never be considered as monochromatic: radiation always contains components whose frequencies occupy a small spectral interval, an interval which can be very small, but not be strictly zero. This is a fact on which Planck had insisted a great deal in his statement of his theory of radiation. Therefore, the relation of Einstein which puts the energy of a corpuscle equal to the product of the frequency of the corresponding classical wave by h has something paradoxical in it, since it makes a well-defined quantity equal to one which is not. The development of wave mechanics has later shown what the real meaning of this difficulty was.

In résumé, the hypothesis of photons, marvelously adapted to the interpretation of the photoelectric effect and the Compton effect, can not lead to a purely corpuscular theory of radiation. It calls for the development of a more comprehensive theory which would attribute to radiation an aspect at once corpuscular and undulatory, these two aspects being related to each other by the Einstein relation. We are going to examine how wave mechanics has tried to reconcile these two aspects and in what measure it has succeeded.

5. The First Applications of the Quantum Hypothesis

THE QUANTUM HYPOTHESIS, strongly confirmed by the success of Planck's theory of black body radiation, and by that of Einstein's theory of the photoelectric effect, was not slow in showing its fruitfulness in many fields. We shall give several examples.

Statistical mechanics, as we have seen, leads to a result known under the name of the theorem of the equipartition of energy. This theorem can be stated in a general fashion by saying: in a mechanical system with a great number of parts, which is found to be in thermal equilibrium at a uniform temperature, the energy of thermal agitation is divided equally among the different degrees of freedom of the system. This theorem which is rigorously deduced from the principles of classical statistical mechanics, is often found to be very well verified: thus it leads to an exact prediction of the mean kinetic energies of the atoms or molecules in a gas and to a generally correct evaluation of the specific heats of these bodies. Nevertheless, the development of the quantum theory has shown that this theorem is not valid in a general way, for it is this theorem which led to the inexact law of Rayleigh-Jeans for the spectral density of black body radiation. Planck's quantum hypothesis had precisely this goal of avoiding an equipartition of energy. If the views of Planck are correct, we then should expect to find some differences with respect to the classical laws in fields other than black body radiation.

Let us take as an example the theory of solid bodies. In a homogeneous solid body, the atoms have their positions of equilibrium where they would rest immobile if there were no thermal agitation. As a consequence of the thermal agitation, the atoms oscillate about their positions of equilibrium with increasing intensity as the temperature is raised. According to the theorem of the equipartition of energy, all the atoms of the solid body ought to have the same mean energy, and the calculation of this mean energy allowed the older statistical mechanics to deduce the following simple and general result: "the specific atomic heat of any solid body (that is, the quantity of heat that it is necessary to furnish to a gram-atom of the solid in order to raise its temperature one degree) is equal to about 6 calories." This is the law of Dulong and Petit which was discovered experimentally by the two physicists whose name it bears before being justified theoretically. This law has been shown to be so accurate for a large number of solid bodies at ordinary temperatures, that chemists, by assuming its validity, have used it in certain cases in order to fix the value of certain molecular weights. But, if the law of Dulong and Petit is often verified, it is not however always so. Certain solid bodies, generally very hard ones, such as diamond, have a specific atomic heat appreciably lower than 6, and for all solid bodies, if the temperature is lowered, there comes a moment when the law of Dulong and Petit is no longer accurate, the specific heat becoming lower than the value which it predicts. The quantum theory has explained these anomalies very well. The atoms of a solid body do indeed vibrate about their positions of equilibrium with a frequency which depends on their mass and on the intensity of the restoring force. According to the quantum hypothesis, the energy of oscillation of the atom ought to be equal to at least a quantum of energy corresponding to the frequency of oscillation. If the thermal agitation can only with difficulty furnish the atom the quantum of which it has need in order to vibrate, the atom will remain immobile and equipartition will not take place. The quantum of oscillation for the atoms of a great number of solids is small enough so that the thermal agitation at ordinary temperatures can easily furnish it to the atom. The equipartition takes place and the law of Dulong and Petit is valid. However, for very hard bodies, as diamond, where the atoms are very firmly held in their position of equilibrium, the quantum of oscillation is so large that the equipartition can not be established at ordinary temperatures, whence results the departure with respect to the law of Dulong and Petit. Finally, as the temperature is lowered there will come a point for all solid bodies when the thermal agitation is no longer sufficient to furnish all the atoms their quantum of oscillation and thus, as a consequence, the specific heat falls below its normal value.

The theory of specific heats, founded on the quantum hypothesis, and accounting for both the successes and failures of the law of Dulong and Petit, was considered first by Einstein, then developed by Nernst and Lindemann and later by Debye, Born and von Karman. It accounts very well for the general trend of phenomena. The quantum theory of specific heats can moreover also be applied, *mutatis mutandis*, to the specific heat of gases: in particular it explains why the degrees of internal freedom of complex gaseous molecules seem to become "ankylosed" at low temperatures, a fact which was inexplicable in classical statistical mechanics.

The quantum hypothesis proved to be strongly confirmed in its first applications. It has found a support equally in the calculation of the upper limit of the continuous spectrum of X-radiation emitted by an anticathode under the impact of electrons of a given velocity. Each of these applications of the hypothesis leads to formulae where the constant h appears, in such a way that by comparing these formulae with the experimental results a value of h can be derived. The values of h thus obtained from the study of very different phenomena are in remarkable agreement.

By 1913 the brilliant and strange conception of Planck was therefore found to be buttressed by numerous facts. Bohr's theory of the atom came out at this moment to bring it a new and striking confirmation and to show at what point the very structure of matter was determined by the existence of quanta.

Chapter 6 | The Atom of Bohr

1. Spectra and Spectral Lines

We can not directly explore the interior of the atom, this unimaginably small microcosm where all the quantities are minute fractions of those that we can perceive. The structure of the atom can be revealed to us only by phenomena observable on our scale which are a consequence of this structure. Among these phenomena are the spectra of the light rays which are emitted in certain conditions of thermal or electric agitation by the atoms of the elements. These light rays are indeed characteristic of the atoms which emit them; they correspond to events which take place in the interior of the atoms and can as a consequence teach us something of their structure. Hence, the study and methodic classification of the spectral lines was a work of major importance in physics.

This work, besides, was not very easy, for light spectra are very complex and, if it is desired to extend their study beyond the visible limits into the infrared and ultraviolet, it is necessary to use special experimental techniques which were made available only by degrees. Nevertheless, in the complexity of the spectra it has been progressively possible to discern certain regularities, to verify certain empiric laws and thus to establish a little order in the very extensive body of experimental results. It was noticed first of all that it was possible to separate the lines into families, into series, to use the technical word, whose structure exhibits great analogies for the different elements. In any one series the different lines have frequencies which exhibit among themselves regular relations capable of being simply expressed by a mathematical formula. So it was that in 1885 Balmer was able to find a formula representing, as a function of a single integer varying from one line to the next, all the frequencies of the lines forming the visible spectrum of atomic hydrogen and forming the series called since then the Balmer series. The exploration of the spectrum of hydrogen beyond the visible limits revealed the existence of an ultraviolet series (Lyman series) and the infrared series (series of Paschen, Brackett and Pfund); in each of these series, the frequencies of the lines obey laws quite analogous to the Balmer law. Spectral series of a similar type, although more complex, are found for elements other than hydrogen, principally for the alkalines. In each series, the frequencies of the lines are always given by formulae of the same type as the Balmer formula, that is, they are expressed as a difference of two terms, the one fixed and characteristic of the series, and the other variable from one line to the next. Due to this particular mathematical expression of the frequencies of the spectral lines it quite frequently happens that the frequency of a spectral line is the sum of the frequencies of two other spectral lines. This group of rules established empirically by the study of the spectra of the elements allowed Ritz to set up a general law that is called "the principle of combination" and which forms the keystone of all of contemporary spectroscopy.

The principle of combination can be stated by saying: "For each kind of atom, it is possible to find a sequence of numbers, called the spectral terms of the given atom, so that the frequency of each spectral line of this atom is equal to the difference of two of these spectral terms." The form of the Balmer law and analogous laws, the fact that between the frequencies there exists relations of addition, all this is immediately understood when the principle of combination is assumed. The validity of the principle of combination is accordingly found to be established in an inescapable fashion by innumerable spectroscopic facts. But this principle should have its *raison d'être* in the structure of the atoms: it should, properly interpreted, furnish us an essential indication on the manner in which the spectral lines are emitted through internal modifications of the atomic structure. Theoretical physics therefore found itself facing an urgent and important task: to explain the origin of the principle of Ritz, and to deduce from it information on the structure of atoms.

Unfortunately the classical ideas of theoretical physics seemed quite incapable of explaining the spectral laws which the experimentalists had patiently succeeded in extracting from the observed facts. In order to explain the emission of spectral lines, the electromagnetic theory invoked in effect the existence in radiant matter of electrified corpuscles in vibration. It imagined, for example, that the atoms contained electrons which normally would be immobile in a position of equilibrium, but which, subjected to certain excitations, would be capable of vibrating periodically around this position. But the laws that could thus be predicted for the distribution of the spectral lines in the scale of frequencies were very different from the real laws. It was this failure of the classical conceptions that Henri Poincaré noted in 1905 when he wrote: "Our first glance at the distribution of the lines makes us think of the harmonics that are met with in acoustics, but the difference is great; not only are the wave numbers not successive multiples of the same number, but we do not find anything analogous to the roots of those transcendental equations to which we are so often led in physical mathematics; such as the one for the elastic vibrations of a body of a certain form, or the one for the Hertzian oscillations in an exciter of a certain form, or the problem of Fourier for the cooling of a solid body. The laws are simpler, but they are of an entirely different nature . . . Of that, we have not taken account, and *I believe that there is one of the most important secrets of nature*." [1]

And I believe that there is one of the most important secrets of nature! A phrase which truly seems prophetic when one thinks that it was written ten years before the theory of Bohr! For it is just the theory of Bohr which furnished the true significance of the spectral laws and showed how these laws imply the quantized character of material structures. It has revealed to us that the entire internal organization of matter and the stability of this organization rests on the existence of quanta. Without quanta, matter could not exist. There is the great secret of which Poincaré spoke.

2. *The Theory of Bohr*

AND so we are led to speak of that famous quantum theory of the atom which Bohr first developed in 1913. We have seen that at that time physicists were inclined to assume a planetary model in which the atom would be constituted by a central nucleus with a positive charge and with a mass almost equal to the total mass of the atom and by electron-planets gravitating around this nucleus. This model, first suggested by Jean Perrin, was found to be strongly confirmed by the experiments of Lord Rutherford and his co-workers, experiments from which it fol-

[1] La Valeur de la Science, p. 305.

lowed that there existed in the atom a nucleus, almost point size and charged electrically. Unfortunately this planetary model, although strongly suggested by experiment, did not fit in at all with the classical ideas concerning the emission of radiation and the motion of electrified particles. Indeed, the fundamental fact that there exists in spectra lines which are sensibly monochromatic and of invariable frequency had necessarily led physicists, imbued with classical ideas, to imagine that in the atoms there were electrified particles, the electrons, having in their normal state a position of stable equilibrium toward which they were restored if they were pushed aside. The electron displaced from its position of equilibrium by any external cause whatever will vibrate around this position with a given frequency and, according to the electromagnetic theory of emission, it will be the origin of a divergent electromagnetic wave of well-defined frequency, since, by thus losing little by little its energy in the form of radiation, it will finally return to rest in its position of equilibrium. Hence, in this way both the monochromatic character of the spectral lines and the stability of the atomic structure will be explained. But the planetary model of the atom did not permit an explanation of this kind at all, for the electrons, describing there their Keplerian orbits, have a frequency of revolution which depends on their energy and varies with it; therefore if the classical theory of the emission of radiation is applicable to the atom, the planetary electrons should progressively lose their energy by giving up radiation of a continually variable frequency and finally fall into the nucleus, neutralizing it. Classical theory, applied to the planetary model, does not let us recover either the monochromatic character of the spectral lines, or the stability of the atom. Such then was the difficulty which confronted Niels Bohr when he began his researches.

Bohr had the great merit of clearly seeing that it was necessary to adopt the planetary model, but to introduce into it the fundamental ideas of the quantum theory. These ideas lead, we know, to supposing that among the infinity of possible motions predicted by classical mechanics, only certain ones, the quantized motions, are stable and normally realized in nature. Planck had arrived, as we have seen, at a general statement defining the quantized motions of a periodic system defined by a single variable. At the time when Bohr wrote his first memoir, it was not known how to quantize periodic motions defined by more than one variable, but it already seemed very probable that the manner of effecting this quantization in the general case would be found. Therefore Bohr was able to suppose that the motion of the atomic systems was quantized. Hence it follows from this that all atoms possess a series of stable quantized states or stationary states and it can be supposed that an atom is always found in one of these stable states. Since an isolated atom forms a conservative system, each of these stationary states will be characterized by a "quantized" value of the energy, and each kind of atom will possess a sequence of quantized values of its energy corresponding to the different possible stationary states. To the atom of each element there will thus be attached a sequence of numbers giving the energies of the different quantized structures of which it is capable.

When we have come to this point in the line of reasoning, we see that the result thus obtained presents a good analogy with the existence of the spectral terms such as they arise from the principle of combination. In order to obtain a quantum interpretation of the spectral terms and of the law of Ritz, it is indeed sufficient to suppose that the frequencies of the spectral lines of the atom are always proportional to the difference of two of the quantized values of its energy. Now Bohr had clearly seen that this postulate presented itself very naturally in the quantum theory of the atom. Indeed, since the quantized states of the atom are stable, the existence of one of these states should not be accompanied by any radiation: this conclusion is evidently contrary to the predictions of the electromagnetic theory since, in the quantized states, the electron-planets are describing closed orbits and constantly undergoing great accelerations, but it is consistent with the idea of quantum stability. Thus it is that when an atom passes from one quantized state to another with a change of energy, that the emission of spectral lines is produced. Bohr therefore assumed that each spectral emission has its origin in the brusque transition of an atom from one stationary state to another with a loss of energy in the form of radiation. In the quantum theory, besides, it is very natural to suppose that energy is emitted by quanta, by photons. Hence, at the moment of a transition, there is emission of a quantum of radiant energy equal to the difference between the energy of the initial stationary state of the atom and the energy of its final stationary state. Whereupon the following proposition known under the name of "Bohr's law of frequencies" immediately results: The frequency of a spectral line emitted during the transition which an atom makes in passing from a stationary state A to a stationary state B is equal to the quotient of the difference between the energy of state A and that of state B by Planck's constant h. According to the law of frequencies, the spectral terms of an atom are equal to the energies of the stationary states of this atom divided by h, and thus the principle of combination finds an interpretation.

In résumé, Bohr based his quantum theory of the planetary atom on the following two bases: (1) the atom possesses a series of stationary states, the only ones which can be physically realized, corresponding to the quantized motions and which can be calculated by Planck's method; (2) the spectral lines of the atom are emitted when the atom undergoes a transition between stationary states, their frequencies being determined by the law of frequencies. The principal work to be done then was to calculate the energy of the stationary states for different atoms. The simplest case is evidently that of hydrogen, whose atomic number is 1. In this case there is only a single planetary electron which must describe a Keplerian trajectory around the nucleus. But, even in this simple case, Bohr could not, at the time of his first efforts, completely treat the problem before him. In order to define a Keplerian motion, two variables actually are needed, for example, the radius vector and the azimuth of the planet. Now it was not yet known how to quantize motions except those defined by a single variable. Bohr solved this difficulty by considering only circular Keplerian motions for which, the radius vector remaining constant, the azimuth can be considered the sole variable. Then, by writing that for the stationary circular orbits the cyclic integral of action is equal to a whole multiple of the constant h, Bohr obtained the expression for the energy of these stable orbits as a function of an integer varying from 1 to infinity. By dividing the expression of these energies by h, the spectral terms of hydrogen were obtained and consequently the formulae representing the frequencies of the different spectral series. Thus the Balmer formula and the analogous formulae for the Lyman, Paschen, etc., series were immediately found without modification. And not only was the form of these formulae found but they were found numerically. In the Balmer formula and the analogous formulae there figures in fact a constant which the spectroscopists had named the Rydberg constant and whose value had been measured very accurately for a long time. Now the theory of Bohr leads to attributing to this constant a value which is expressed with the aid of the fundamental constants: charge and mass of the electron and Planck's constant. Therefore the theory of Bohr allows us to calculate *a priori* the Rydberg constant and this calculation furnishes exactly the value previously measured by the spectroscopists. This quantitative concordance has been a great success for the atomic theory of Bohr. It has proved that the way opened up by Bohr was the right one.

Bohr was not content with this first remarkable success. He extended his theory to the case of ionized helium. Helium is the second element in the Mendelejeff table where the elements are arranged in order of increasing atomic weights: its atomic number is 2 and according to the planetary model, the helium atom should be formed by a nucleus whose electric charge is double that of the proton and by two planetary electrons. The mathematical problem of the determination of the quantized motions in the helium atom is therefore complicated for it is a mechanical problem of three bodies. But it is simplified if, in consequence of an external action, the helium atom loses one of its electrons. We then have a singly ionized helium atom where there is only one electron and we have the same mechanical problem as in the case of the hydrogen atom, with this sole difference that here the electric charge of the nucleus is twice as great. Bohr then showed that the spectral lines of ionized helium should obey laws quite similar to that of Balmer, but in which the Rydberg constant is multiplied by 4. This led his attribute to ionized helium, the Pickering series, discovered in the spectra of certain stars and which, up to then, had wrongly been attributed to hydrogen. This application of the quantum theory of the atom has thus aided in unraveling a mass of spectroscopic facts whose interpretation had been doubtful.

Moreover, Bohr was then able to explain a little fact which, at first, had seemed singular enough. Experiment indicates that for the spectrum of ionized helium, the Rydberg constant (corrected for the factor of 4 of which we have just spoken) does not have exactly the same value as for the spectrum of hydrogen. Bohr saw the origin of this discrepancy in the fact that the nucleus of the atom undergoes a reaction from the planetary electron and consequently does not remain rigorously immobile. The original theory, in which he had considered the nucleus as the fixed center of attraction, is then only a first approximation: it was advisable to take account of the motion of the nucleus which is that much more important when the nucleus is lighter. By returning again to his calculations in a more rigorous fashion, Bohr found a corrective term depending on the ratio of the mass of the electron to that of the nucleus. Since the nucleus of helium is about four times heavier than that of hydrogen, the corrective term thus calculated is therefore sensibly greater for hydrogen than for helium, both being still very small. This then explains why the Rydberg constant does not have exactly the same value for the two substances and the difference predicted by the calculation of Bohr coincides exactly with the data of experiment.

The theory of the Bohr atom also leads to an understanding, at least in a general fashion, of the structure of the optical spectra of elements other than hydrogen and ionized helium. Assuredly, when we seek to extend Bohr's method of calculation to atoms with more than one electron, we run headlong into great difficulties: on one hand the problem becomes complicated or even unsolvable, on the other hand the application of the rule of quantization becomes uncertain. Nevertheless, the general analogy of the spectra of all the elements and the appearance in their series formulae of the Rydberg constant shows the profound relationship of all the spectra and makes us believe that the method, crowned with success in the case of hydrogen, should be able to be taken over for other elements. We can with Bohr adopt the following scheme, which is certainly very rough: consider a non-ionized atom of atomic number N as containing in its central region, around the nucleus, N-1 electrons in motion; the Nth electron is supposed to be moving around this "electronic carcass" and it is its transitions from one stable state to another which determine the spectrum of the element. As a first approximation, the action of the nucleus and of the carcass is reduced to that of a Coulomb field and spectral terms analogous to those of hydrogen are found. Thus we find an interpretation, rough enough to be sure, of the analogy of all optical spectra.

By following this same line of ideas, we also come to understand the nature of the spectra of the X-rays which present in the main the same general characteristic as the optical spectra. We do not wish to enter here into great detail on this subject and we shall say only that the ideas of Bohr lead us to an understanding of the origin of the great law of X-ray spectroscopy: the law of Mosely. Just as optical lines, the lines of the spectra of the Röntgen rays are divided into series which are found to have the same general structure for all elements. When the discovery of the diffraction of X-rays by crystals due to von Laue, Friedrich and Knipping (1912) allowed us to measure exactly the wave length of X-rays, a young English scientist, Moseley, saw that if the homologous lines are followed through the spectra of different elements, a displacement of the lines in the scale of frequencies is observed which is almost proportional to the square of the atomic number; in other words, the frequency of a certain line is about four times smaller in the spectrum of a certain element than in the spectrum of an element of twice the atomic weight. The formulae of the Bohr theory easily show that the frequencies of all the spectral lines in the X-ray region should vary from one element to another just about as the square of the atomic number, at least as a first and very rough approximation. The law of Moseley thus is found justified and in this manner the atomic theory of Bohr has shown its heuristic power in all the spectral regions.

3. *The Perfecting of the Theory of Bohr. The Theory of Sommerfeld*

FROM the point of view of its mathematical development, the theory of Bohr has a grave gap. Even in the simplest case, that of the hydrogen atom, it allows us to calculate only the quantized energies of the circular trajectories and does not deal with elliptical trajectories. The reason for this inability is found in the insufficient development of the methods of quantization. The method of quantization of action given by Planck is applicable actually only to motions for the description of which a single variable is sufficient. In order to be able to develop the atomic theory of Bohr in all its fullness, it was indispensable that the following problem be resolved: to find a method of quantization applicable to mechanical systems with more than one degree of freedom.

This problem was resolved almost simultaneously in 1916 by M. W. Wilson and by Sommerfeld. They took note of the fact that the mechanical systems in which the quantum theory was interested all belonged to the category of quasi-periodic systems with separable variables. For systems of this kind, the different variables vary periodically, but their periods are in general different. Moreover, it is possible, if the variables have been properly chosen, to break down the integral of action into several integrals, each of which depends only on a single variable. By extending each of these integrals to a complete cycle of the corresponding variable, there is obtained a quantity called "cyclic period of the integral of action" and there are evidently as many of these periods as there are variables. It suffices then to write that each of these cyclic periods is a whole multiple of the constant h in order to obtain a general statement sufficient for the quantization of the motions of the systems. In particular, if there is only one variable the statement of Planck is recovered.

The Wilson-Sommerfeld method of quantization of which we have just sketched the over-all picture allows us to solve in principle all the problems that the atomic theory of Bohr had put to itself. Naturally, in practice, when it is a question of atoms even just a little complicated, the mechanical problem becomes inextricable and we are stopped, but this check is connected with the impossibility of solving the dynamic equations, and no longer with the imperfection of the method of quantization.

Sommerfeld used this method of quantization, of which he was one of the inventors, in order to treat various problems of the atomic theory that Bohr had been unable to do. He had first shown that for the hydrogen atom the consideration of elliptical orbits does not introduce any new values in the sequence of the quantized energies and consequently modifies nothing in the original conclusions of Bohr. Concerning optical spectra, he was able to show that by taking account, a little schematically it is true, of the overlapping of the electron trajectories, other formulae, up till then empirical, could be found in place of the laws of the Balmer type, which were known to spectroscopists under the name of the Rydberg and Ritz formulae, and which represented better than the formulae of the Balmer type the exact distribution of the optical lines in the scale of frequencies.

But the great success won by Sommerfeld was his theory of the fine structure. A detailed study of the spectrum of hydrogen, made with the aid of a spectroscope of high dispersive power, had shown, in point of fact, that certain of the lines of the spectrum of hydrogen were not simple, but were composed in reality of several lines with very close frequencies. The formulae of the Balmer type, found in the theory of Bohr, did not take account of this fine structure. Sommerfeld had the very ingenious idea of investigating if this complexity of the spectral lines could not be interpreted by applying to the intra-atomic electrons, in place of the classical Newtonian mechanics, the Einsteinian mechanics of relativity. In fact, if we look again at the formulae of the Bohr theory, we see that, in the atoms, the electrons, according to the planetary scheme, possess velocities sufficiently high so that it is advisable to take account of the relativistic corrections. After re-doing these calculations, using both the method of quantization and the dynamics of Einstein, Sommerfeld found that certain of the quantized values of the energy predicted by the former theory were split; in other words, certain of the spectral terms of hydrogen predicted by Bohr were broken down into spectral terms with values very close together. Obviously this was sufficient to explain the phenomenon of the fine structure, and the values calculated by Sommerfeld for the difference of frequency between the components of the doublets of the Balmer series were found to be in very good accord with the experimental values.

Encouraged by this success, Sommerfeld wished to explain, in the same way, the fine structures observed in X-ray spectra, fine structures which are even more important than those of the optical spectra. In fact there are observed in X-ray spectra doublets whose components are easily separable and whose variation can easily be followed all along the series of elements. Certain of these doublets, called "regular doublets" show a difference of frequency which increases rapidly with the atomic number of the element, about as the fourth power of the atomic number. The union of the dynamics of relativity with quantum methods permitted Sommerfeld to interpret the existence of these doublets and their variation as N^4. In particular, the doublets of the L series are very well represented by the formulae of Sommerfeld.

The very good results thus obtained by Sommerfeld had seemed with their publication (1916) to constitute a magnificent and definitive success both for quantum methods and for relativity mechanics, and had called forth a legitimate enthusiasm. But a more detailed examination was not long in showing that several shadows remained on the picture. First the body of methods and conceptions utilized by Bohr and Sommerfeld, which today constitute the "old quantum theory," raised some difficulties of principle to which we shall return in the last section of this chapter. But besides these difficulties of a general nature, the results of Sommerfeld ran into objections of a more particular sort. In the first place, the actual fine structure of the optical spectra and the Röntgen spectra is more complex than the theory of Sommerfeld would indicate. The scheme of spectral terms established by Sommerfeld, although more complete than that of Bohr, is not yet as rich as that whose existence is proven by spectroscopic experiments. The difficulty is grave, for the quantum method of Sommerfeld leaves no place for the introduction of the supernumerary spectral terms revealed by experiment: a homogeneous and complete method, it does not seem capable of being enlarged. Sommerfeld had succeeded in classifying the supernumerary spectral terms by introducing a supplementary quantum number that he called the "internal quantum number," but rationally there is no justification for introducing this new strange element into the principles of the theory. The much more recent discovery of the magnetic character of the electron was necessary in order to justify and explain the appearance of the internal quantum number.

The theory of Sommerfeld was thus too narrow to be able to explain completely the fine structure of spectra. At least, it would seem to predict very exactly the doublets of the Balmer series and those of the X-ray spectra. Unfortunately, a close examination of the structure of spectra has not subsequently entirely confirmed this favorable impression. This examination has led, indeed, to characterize each of the stable states of the atom by a certain group of quantum numbers and this distribution of the quantum numbers seems very certain. Now, if these are taken into account, we come to the following singular conclusion: the theory of Sommerfeld correctly predicts the existence of the doublets of the Balmer series and of the X-ray spectra, but it does not put them where they really are. It is scarcely possible to attribute to chance the apparent success of the formulae of Sommerfeld, but we feel that in the theoretical edifice something is not yet in its proper place. It is the theory of Dirac, who, by combining wave mechanics and the magnetic nature of the electron, has put things in their proper place, still conserving the essential results of Sommerfeld. Thus it has seemed that the guiding ideas of this eminent physicist were correct, but at the time when he wrote his theory, the quantum doctrine on one hand, and our knowledge of the electron on the other, were not far enough advanced to permit him to make an entirely satisfactory work.

4. The Theory of Bohr and the Structure of Atoms

THE ESSENTIAL IDEA of the theory of Bohr is that within the atom, the electrons can be found only in certain stationary states of quantized energy. Thus there exist levels of energy among which the different electrons are distributed. Now we know that there exist 92 elements for which the number of electrons contained in the atom increases regularly from 1 to

92. If then we consider all the elements successively in order of increasing atomic number, we shall see their internal electronic organization become progressively more complicated by the addition, one by one, of new electrons, and by thus following the evolution of the internal structure of the elements, we should in principle be able to account for their chemical, spectroscopic and even magnetic properties. Before the birth of the quantum theories, the Russian chemist, Mendelejeff, had arranged all the elements known in his day into a list which corresponds to the order of increasing atomic weights, that is, almost exactly to the order of increasing atomic numbers. It was then seen that a certain periodicity of the chemical properties of the elements thus classified existed, that is, there reappeared in this list, at regular intervals, substances having similar chemical properties. In reality, this periodicity is not of a very simple nature: the periods are shorter at the beginning than at the end of the Mendelejeff table and, here and there, certain mishaps occur which disturb the regularity. Nevertheless, the existence of a periodicity is incontestable, and a good theory of the atom should account for this. The theory of Bohr, to accomplish this mission, laid down in principle a rule whose profound significance we shall see better later on. It assumed that at each quantized level there could not be more than a maximum number of electrons. In other words, the intra-atomic energy levels are susceptible of being saturated with electrons. This was truly an entirely new and unforeseen property of the quantized structures that had been assumed a little surreptitiously and without anyone being aware of its importance.

The postulate of the saturation of the levels being assumed, it is easy to understand the nature of the periodicity observed in the Mendelejeff table by relying on the great principle of physics according to which the stable state of a system is always a state of minimum energy. If the saturation of levels did not exist, for all the elements the totality of the electrons in the normal stable state of the atom would be found on the level of least energy.

But because of the saturation of the levels, things are otherwise: when we pass from one element to the following, the additional electron which is added to the structure of the normal atom is placed on the level of least energy which is not yet saturated, or as it is often said, on the level of least energy where there is still a free place. When, for an element, the lowest level is found to be saturated with electrons, the additional electron of the following element must then find its place on the level which comes next in the order of increasing energies. Thus if one follows the development of the structure of the atom along the Mendelejeff table, one should see the various lower levels of the atom fill up and become saturated one after the other. But here an important observation should be made: the existence of the fine structures has taught us that the quantized levels of energy for an electron within an atom are distributed into groups in which the energies are very close together. We shall say that the electrons with nearly the same energy which are on the levels of the same group form a "shell." Since the levels are saturated one after the other when one follows the successive building-up of the elements, the different shells are progressively formed. The steps in the building-up of one shell correspond to a sequence of well-defined chemical or spectroscopic properties; then, when the shell is saturated, the building-up of the following shell is begun where almost the same steps are repeated. Thus the periodicities of the observed properties along the table of elements finds an entirely natural explanation. The fact that the different shells contain different numbers of levels and require for their saturation different numbers of electrons, implies the existence of differences in the length of the periods in the Mendelejeff table. We shall limit ourselves here to these brief indications. The interpretation of the variations of the properties of the element as a function of the progressive complication of their electronic structure was first proposed by Kossel; it has subsequently been developed and deepened by the works of Bohr, Stoner and Main Smith and forms today a satisfactory body of work.

The distribution of electrons between the shells and the levels is intimately related to the structure of the X-ray spectra. According to the theory of Bohr, the origin of X-rays is, in effect, the following: if an external action detaches from an atom an electron situated on one of the lower shells, there is consequently in this shell a free place and one of the electrons situated on one of the outer shells will be able to come and take its place by losing its energy, the transition thus realized giving place, according to the fundamental ideas of Bohr, to the emission of a quantum of radiation. The radiations thus emitted are the lines of the X-ray spectra. It can therefore be understood, without our insisting here on the details, that the study and classification of the spectral lines in the Röntgen region have contributed greatly to rendering precise our ideas on the internal structure of the atoms and on the saturation of the levels. It can be said that the phenomenon of the saturation of the levels, whose importance we have underscored, is absolutely proven by the progressive development of the X-ray spectra along the table of elements.

The ideas of Bohr on the existence of the quantized levels in the atoms, as well as the schematic pictures of the atomic structure of the different atoms, have been strongly confirmed by experiments on ionization by collision. Ionization by collision consists in the detachment of an electron from an atom as the result of a collision. The lower the level on which an electron is situated the more energy is required in order to detach it from the atom. Let us imagine that a beam of particles of known energy is directed toward the atoms of a gas. The collision of the particles with the atoms of the gas will be able to bring about the detachment of electrons within the atoms for which the energy of detachment is lower than that of the incident particles. If we progressively increase the energy of the particles which are used in the bombardment, we shall then see a new kind of ionization appear each time that this energy exceeds the value corresponding to the detachment of an electron from one of the levels of the bombarded atom. The successive appearance of new kinds of ionization will thus furnish us, at least in principle, with the complete schematic of the levels in question. Experiments of this kind, of which Franck and Hertz were the initiators, have entirely confirmed, in perfect accord with the indications furnished by the Röntgen spectra, not only the existence of the quantized levels, but the expected distribution of the different levels in the different atoms.

5. *Criticism of the Theory of Bohr*

WHAT WE HAVE SAID in this chapter will suffice to show the importance of the atomic theory of Bohr. Its birth marked an important step in the history of contemporary physics: with its appearance it permitted a unification of the immense domain of spectroscopy, and made the character of the laws which work in it understandable; since then, generalized under the form of a coherent theory of quantization (called today the old quantum theory), it has won many victories in the explanation and prediction of atomic phenomena.

However, the admirable body of doctrine founded on the ideas of Bohr was not immune to criticism. We do not even mean those few failures which it experienced here and there; e.g., difficulties we have already noted in adapting the formulae of the fine structure due to Sommerfeld to the spectroscopic facts, or the result in disagreement with experiment which Kramers obtained, after long calculations, when he wished to apply the methods of the old quantum theory to the theoretical evaluation of the ionization potential of the neutral helium atom. These failures in themselves did not auger very well for the future of the theory, but other criticisms of a more general nature could be directed against the original conceptions of Bohr and would show that they could not be considered either coherent and complete, nor as a consequence truly satisfying. We shall say a few words about each of these criticisms.

First, the theory of Bohr was quite incapable of making entirely precise the nature of the radiation emitted during quantum transitions. Assuredly, it gave a precise rule for calculating the frequency of this radiation, but, in order to have a complete description of even monochromatic radiation, it is necessary to know further its intensity and its state of polarization. A great deal less precise from this point of view than the classical electromagnetic theory of radiation, the original theory of Bohr gave no indication of the intensities or the polarization. Bohr was acutely aware of this defect in his theory and it was he who first, in 1916, sought to remedy it by formulating his principle of correspondence. Since the following chapter will be devoted to this important question, we shall not dwell on it at this moment. But, besides this lack of precision concerning the emission of radiation, the theory of Bohr had other weaknesses. In particular, it rested on a very strange alliance of conceptions and formulae from classical dynamics on one hand, and of quantum methods on the other hand. In it one started by comparing the intra-atomic electron to a material point of classical mechanics describing its orbit in a very regular way under the influence of Coulomb forces, taking as a picture of the atom a little planetary system of extraordinarily reduced dimensions. Then, in this representation completely consistent with classical ideas, there is brusquely introduced in an offhand way the conditions of quantization, by asserting that, among the infinity of trajectories predicted by the calculations of classical dynamics, only those were stable and physically realizable which complied with the demands of the quantization. Thereupon, the changes in the state of the atom were able to consist of nothing more than brusque transitions accompanied by a loss of energy by radiation and there seemed no possible way permitting us to describe these brusque transitions in the classical framework of space and time. Between transitions, the atom is in a stable state, one of the "stationary states" of Bohr, where it seems to ignore completely the external world, for it radiates no electromagnetic energy, despite the precise prescriptions of the electromagnetic theory; then suddenly, it jumps from this stationary state to another by effecting a transition impossible to describe or represent in space. We are now very far from the classical conceptions, after having taken them for our point of departure, and a theory which starts out from a system of concepts and ends by completely denying it, is evidently scarcely coherent. And all the dynamic imagery which has been introduced in the beginning of the theory, these point-like electrons describing orbits of a perfectly calculable form along which they have at each instant a well-defined position and velocity, this all finally served only to calculate the energy of the stationary states and the corresponding spectral terms which, alone, can be compared with experiment, thanks to spectroscopic measurements and the results of ionization by collision. Is one not tempted to think that all this over-precise representation is artificial, that the forms of the orbits and the values of position and velocity for the electrons do not correspond to any physical realities, and that only the values of the energies of the stationary states finally furnished by all this quantized celestial mechanics have a real physical meaning?

As often happens, the gifted inventor of the quantum theory of the atom was the first to see and underscore the weaknesses of it. He was the first to insist on the fictitious character of the planetary model, on the entirely new nature of the conceptions of the stationary states, on the impossibility of making these conceptions fit into the ordinary framework of space and time, on the necessity of opening up radically different paths. By his principle of correspondence he indicated one of the directions to follow, and guided by these ideas, his pupil, Werner Heisenberg, succeeded several years later (in a remarkable and very original effort of which we shall speak again) in creating one of the aspects of the new quantum theory: quantum mechanics.

Chapter 7 The Correspondence Principle

1. The Difficulty of Integrating the Quantum Theory with That of Radiation

THE ELECTROMAGNETIC THEORY, completed by the electron hypothesis, gave a perfectly clear and precise picture of the mechanism of the emission of radiation by a system of charges in motion. The structure and motion of a group of electric charges being given, it permitted us to calculate exactly the frequencies, intensities and polarization of the emitted radiation. To succeed in this, it proceeded in the following fashion. First, it calculated in a system of rectangular axes the components of a vectorial quantity, the electric moment of the system, which is defined at each instant by the positions of all the charges constituting the system. These components are functions of the time, which, according to the general theorems of the mathematical theory of Fourier's development in series or integrals, can be developed in a sequence (finite or infinite) of harmonic terms. The electromagnetic theory tells us that the system will emit radiations possessing all the frequencies which figure in the Fourier development. Moreover, a radiation having one of these frequencies and whose electric vector is parallel to one of the rectangular axes should have an intensity which is immediately deduced from the coefficient of the harmonic term with this frequency in the Fourier development of the component of the electric moment parallel to the axis under consideration. These rules suffice to determine completely as to frequency, intensity and polarization, the various radiations emitted by the system under consideration.

If then the electromagnetic theory in the form given it by Lorentz were actually applicable to the elementary particles of electricity, it would allow us to calculate without any ambiguity the radiations emitted by an atom of the Rutherford-Bohr planetary model. We have already seen to what grossly inexact predictions we would be led. If an atom constantly lost its energy in the form of radiation, its electrons would all end by falling very rapidly into the nucleus, and the frequency of the radiation emitted would constantly vary in a continuous fashion. The atom would be unstable and there could not exist spectral lines of well-defined frequencies—which are absurd conclusions. To avoid this essential difficulty, Bohr, as we have seen, had supposed that the atom in its stationary states does not radiate; this is tantamount to denying the possibility of applying the electromagnetic theory of radiation to the orbital motion of the electrons in their stable trajectories.

Having thus severed all relation with the electromagnetic theory, the quantum theory of the atom seemed thoroughly without means to foresee the characteristics of the radiations emitted in the form of spectral lines. However, we have seen how Bohr had resolved the question as far as the frequencies of the spectral lines were concerned, by virtue of the hypothesis that each transition between quantized states is accompanied by the emission of a quantum of radiant energy. But this law of frequencies makes the emitted radiations precise only in a very incomplete way, since it tells us nothing of the intensities and polarizations. Bohr succeeded in 1916 in filling this gap, at least partially, by following a very original, if somewhat disconcerting, method that essentially consists in this: despite the failure of the classical electromagnetic theory in the atomic domain, a method nevertheless is sought to establish a certain correspondence between the quantum phenomena and the formulae of electromagnetism in such a way as to enable us to understand why the electromagnetic theory furnishes a good representation of large scale phenomena. So Bohr succeeded in formulating a very curious "Principle of Correspondence" which has played a considerable and very beneficial role in the evolution of the quantum theory.

Before beginning a study of the Correspondence Principle, we must first closely limit the difficult problem to which Bohr was going to try to give a solution. It is necessary to understand clearly how different are the representations that the classical theory on one hand and the quantum theory on the other had proposed for the phenomenon of emission. In the classical theory, an atomic electron in motion radiates, in a continuous fashion, a whole series of radiations: the emission of these radiations is therefore both continuous and simultaneous. In the quantum theory, on the contrary, an atomic electron does not radiate when it is in a stationary state and when it jumps from one state to another, it emits a single quantum of a monochromatic radiation: the different monochromatic radiations emitted by a group of atoms of the same kind (for example the various spectral lines emitted by a gaseous mass of an element) correspond therefore to transitions undergone by different atoms. In other words, according to the quantum theory, the emission of the spectral lines of an element is discontinuous and proceeds by isolated individual actions. Surely then it is difficult to find two conceptions more different from each other than the classical conception and that of the quantum theory and at the very first it can legitimately be asked if any bridge can be built to connect them.

When we reflect on the means of establishing a correspondence between the classical picture of the emission of the spectral lines and the so dissimilar picture of it that the quantum conceptions suggest to us, we at once perceive that this correspondence, if it ever is realized, can only be of a statistical nature. Indeed, a correspondence with the classical picture can evidently be established only by considering simultaneously the emission of all the spectral lines: now, from a quantum point of view in which the emission of each quantum of monochromatic radiation is an individual act, this is possible only by considering a collection of a very large number of atoms of the same nature, a collection where individual transitions of all sorts are constantly occurring, accompanied by the emission of the various spectral lines of the element under consideration. On the other hand, the indispensable notion of the intensity of the various lines can be introduced into the quantum theory only by treating it likewise from a statistical point of view. The quantized atom, when it undergoes a transition, emits just a single quantum, a single unit, of monochromatic radiation; for such an individual act of emission, there would be no thought of a question of intensity of the radiation. In order to define an intensity, it is thus necessary to consider again a collection of a great number of atoms of the same nature: in this collection a great number of transitions of each kind takes place per second, and, by considering all the transitions of a certain sort and all the quanta of radiation of the same frequency the emission of which accompanies these transitions, an intensity can be defined statistically as the mean density of these quanta in space, and this intensity will be able to be compared with that calculated by the classical theory.

The reader is doubtlessly beginning to perceive how the establishment of the sought-for correspondence might be possible. There will be considered on one hand a collection of fictitious atoms obeying the laws of classical electromagnetism, and on the other a collection of real quantized atoms, and we shall seek to establish a relation between the frequencies, intensities and polarization of the radiations emitted by each of these two collections in such a way that the calculation of the spectral emissions of the first system by the well-known, classical electromagnetic method gives some indication about the spectral emissions of the second, that is about the real emissions. To find the relation to be established assuredly is not *a priori* an easy task. The remarkably penetrating mind of Bohr, however, knew how to accomplish this and to find for this arduous problem, if not a complete and definitive solution, at least a provisional solution which has shown itself to be extremely useful and full of profound physical significance. The time has come to outline it.

2. *The Correspondence Principle of Bohr*

WE SHALL therefore compare a collection of a great number of fictitious atoms which would obey the classical laws, and a collection of the same number of real quantized atoms. If we know the motion of the electrons in the atoms of the first collection, we will know how to calculate the frequencies, intensities and polarizations of the emitted radiations. We wish to derive from this predictions about the frequencies, intensities and polarizations of the radiations emitted by the real atoms. If we knew nothing about these latter quantities, there would be no means of approaching the problem. Fortunately, we know the frequencies emitted by the quantized atoms, thanks to Bohr's rule. Therefore the first thing to do is to compare these frequencies with those that the fictitious atoms would emit according to the classical theory. If this comparison is made, it is seen that in general there is no simple relation between the two categories of frequencies, and we see no means of making further progress on the way that we have undertaken. Here is where the ingeniousness of Bohr asserted itself in a decisive fashion. Bohr had noticed that the electromagnetic theory is always very approximately verified in the domain of macroscopic phenomena. Now, from the quantum point of view, macroscopic phenomena are those which call into play the higher quantum numbers. This makes it probable that the predictions of the quantum theory ought to tend asymptotically toward those of the classical theory in the domain of the large quantum numbers. Therefore it is in this domain that the juncture between the two theories should be effected. And since we know how to calculate both the classical frequencies and the quantum frequencies, the first thing to do is to verify how effectively these frequencies coincide in the case of the stationary states for the higher quantum numbers.

Let us then consider in the quantized atom one of the outermost electronic trajectories corresponding to large values of the quantum numbers and let us simultaneously consider the same electronic trajectory in the fictitious classical atom. In the classical atom, the electron emits continually a whole series of frequencies which moreover are harmonics of a certain number of fundamental frequencies determined by the harmonic analysis of the motion of the electron. In the quantum atom, the electron in stationary motion does not radiate, but it is capable of undergoing transitions which will give rise to radiations whose frequency is definitely determined by Bohr's rule. Then it is seen that to each frequency predicted by the classical theory for the fictitious atom, there corresponds a certain transition of the quantized atom which gives rise to an emission of the *same* frequency. So, in the domain of the large quantum numbers, there is a good coincidence between the frequencies emitted according to the classical mechanism and the frequencies that the quantized electron can emit in undergoing a transition. But, while each classical atom emits continually and simultaneously each of the frequencies in question, the quantized atom can emit only a single one of them by an individual action. But this profound difference between the mechanisms of emission does not prevent the end result from being the same: the two collections of atoms that we are comparing in thought will emit (in the domain of the large quantum numbers) the same spectral lines.

Having thus verified the identity of the predictions of the classical theory and the quantum theory relative to the frequencies in the domain of the large quantum numbers, Bohr considered as certain that, still in this same domain, the predictions of the classical theory relative to the intensities and polarizations of the collection of fictitious atoms would also be exact for the collection of the real atoms. For the real quantized atoms, the emission of the spectral lines is produced by individual transitions between quantized states: as we have explained above, the intensity of a spectral line will then depend on the proportion of the atoms of the collection which will undergo on the average the corresponding transition in a unit of time, that is, on the probability that each quantized atom will undergo the called-for transition in a unit of time. If it is assumed with Bohr that the intensity of the given spectral line emitted by the second collection ought to be equal to the intensity, calculated classically, of the same spectral line for the first collection, this will allow us to evaluate the probability of the quantum transition with the help of the formulae of the electromagnetic theory. Thus we find solved the problem of predicting, at least for the large quantum numbers, the intensities of the spectral lines. What was lacking in the original theory of Bohr in order to make this prediction, was a method of evaluating the probabilities of the quantum transition. The idea of establishing a correspondence between each of these quantum transitions and the harmonic components of the classical radiation led, in the limits of the asymptotic case considered, to a simple and rigorous rule for evaluating the probabilities of a transition. Similarly, it was sufficient to assume, which was quite natural, that the polarizations of the spectral lines really emitted were the same as those predicted by the classical theory in order to solve completely the question of polarization.

Unfortunately this whole remarkable manner of joining images, which were irreconcilable in appearance, in order to fill in the gaps in the quantum theory was valid only in the domain of the large quantum numbers. Now, practically, for a theory of the atom, this domain is the least interesting for, with the exception of certain quite exceptional states of excitation, the atomic electrons are always in stationary states corresponding to the smaller values of the quantum numbers and the usual spectral lines are emitted during transitions between such states. There is then no simple relation between the real quantum frequency and the frequencies predicted by the classical theory, for an atom in a state corresponding to the initial state before transition or the final state after the transition. Nevertheless, Bohr had postulated with great boldness that by extrapolating to the small quantum numbers the correspondence established for the large ones, it should be possible to utilize the classical evaluations of intensities and polarization to predict approximately the real intensities and polarization. We can not explain here in detail how Bohr tried to give a precise form to this "Correspondence Principle." We shall only say that he took a certain average of the classical quantities calculated for a group of states (non-stationary) which are intermediate between the initial stationary state and the final stationary state corresponding to the given spectral line. Although the Correspondence Principle thus formulated has led to interesting and generally exact results, there is the impression that its statement has a somewhat artificial character and that it has not been able to find within the framework of the old quantum theory its definitive formulation. We shall see that it has found in the framework of the new mechanics an expression which is a good deal more thorough. But the importance of the idea presented by Bohr was subsequently shown to be considerable. The idea that the electromagnetic theory, although it was not rigorously exact, was of great value in guiding us, like Ariadne's thread, toward the progressive discovery of the true laws of the elementary quantum theory, proved to be very fruitful. It has served as the basis of a genuine method of correspondence, and the pupils of Bohr, relying on this method and imbued, as Heisenberg has said, with the spirit of Copenhagen, have been able to make progress along this route and by it make some worthy discoveries, as we shall soon relate.

3. Some Applications of the Correspondence Principle

THE Correspondence Principle has let us calculate, at least approximately, the intensities of the various spectral lines, whether in normal spectra, or in spectra modified by the Stark or Zeeman effect. The result of these calculations has been found to be, in general, in satisfactory accord with experiment.

One of the most important applications of these evaluations of intensity has been the investigation of the spectral lines, predicted by Bohr's rule of frequencies, which are emitted with zero intensity, that is, which, in fact are missing in the observable spectrum. It will be useful to explain this point. When all of the stationary states and consequently all of the spectral terms of an atom are known, we immediately obtain all of the spectral lines that the atom can emit by combining the spectral terms two by two according to Bohr's rule. Now, if a tabulation of the lines thus obtained is compared with a list of the lines that actually appear in the spectra, it is seen that all of the predicted lines are not effectively emitted. In other words, all the frequencies of the real spectral lines can be predicted by the combination of spectral terms, but the inverse is not true for all combinations of spectral terms do not always furnish a frequency actually represented in the spectrum. Hence theory ought to be able to furnish "rules of selection" permitting us to say which, among the combinations of spectral terms, are those which correspond to the lines effectively emitted. To do this, the absence in real spectra of certain lines predicted by the combinations is interpreted by supposing that these lines, theoretically existing, are in usual circumstances emitted with a zero intensity. This opinion is confirmed by the fact that in exceptional circumstances, for example, by the action of particularly intense electrical disturbances, the atom begins emitting the lines habitually absent in the spectrum. The Correspondence Principle lets us show that, in usual circumstances, the intensity of spectral lines associated with certain transitions is zero, which means that the probability of an atom undergoing this transition is zero. Thus, among the quantum numbers which define a stable electron trajectory there is one of them called the "azimuthal quantum number." The Correspondence Principle lets us show that only those transitions where this azimuthal quantum number increases or decreases by unity have, in usual circumstances, a non-zero probability of taking place. The following selection rule can be deduced from this: "In usual circumstances, all the spectral lines corresponding to transitions where the azimuthal quantum number does not increase or decrease by unity have a zero intensity and are in fact absent from the spectra." This selection rule, completed by other analogous ones, is remarkably well verified in all of the luminous and Röntgen spectra and considerably facilitates the classification of lines not yet identified. The Correspondence Principle has rendered valuable service by showing the theoretical significance of these selection rules for which an earlier justification had also been proposed by starting out from other considerations. (Rubinivicz)

The phenomenon of the dispersion of light was very difficult to interpret in the quantum theory. Experiment shows, in effect, that the index of refraction varies as a function of the frequency of light and shows large changes in the neighborhood of certain critical frequencies which are exactly equal to the frequencies of the spectral lines which the given substance can emit. The old theories explained these variations of the index of refraction well enough, thus furnishing satisfactory interpretations of the phenomenon of dispersion. The electron theory, in particular, considered the material atoms as containing electric charges capable of vibrating harmonically around a position of equilibrium (electronic oscillators). These charges being capable of giving rise to radiations by means of their vibrations, the frequencies of vibration of the various atomic oscillators should be equal to the frequencies of the spectral lines of the atom. Now, by studying the manner in which a monochromatic light wave falling on an atom sets the oscillators of this atom into forced vibration and the manner in which these forced vibrations of the intra-atomic vibrators react on the propagation of the incident wave, the electron theory succeeded in finding for the variation of the index as a function of the frequency a formula of dispersion thoroughly in accord with experiment: in this formula, the critical frequencies of dispersion were equal to the proper frequencies of the electronic oscillators, that is, to those of the spectral lines of the given substance, a conclusion consistent with the facts. In the Bohr theory, an exact explanation of dispersion was a great deal more difficult. In the Bohr atom, indeed, the mechanical frequencies of revolution of the electrons in their orbits bear no simple relation to the optical frequencies of the spectral lines associated with transitions and not with states. It then becomes very difficult to understand how the change of the mechanical state of the atom by an external light wave can give rise to the phenomenon of dispersion where the principal role is played, not by the mechanical frequencies of the atom, but by the optical frequencies of the spectral lines. This difficulty had not escaped Bohr and his followers. The appearance of the Correspondence Principle had allowed him to seek a solution of it along this new path. Two pupils of Bohr, Kramers and Heisenberg, were able to obtain, during 1923, a quantum formula for dispersion which, without being entirely identical to the classical formula, is thoroughly in accord with experiment. The reasoning of Kramers and Heisenberg was perhaps not absolutely indisputable, but they were constantly being directed and inspired by the spirit of the method of correspondence. As we have said, the formula so obtained was not quite identical with the classical formula: it contained supplementary terms whose real existence was subsequently shown by the experiments of Ladenburg.

In the course of his researches for a formula of dispersion Heisenberg had convinced himself that it would be useful to eliminate as much as possible from the theory of Bohr all those elements that were not directly observable with an increase in use of the observable elements, for example, to cause the frequencies of the electrons in their orbits to disappear with increased use of the spectral frequencies associated with the transitions by Bohr's rule. It is certain that this observation must have contributed to orienting the young scientist on the road that led him, a little later, to the discovery of quantum mechanics.

The quantum theory of dispersion, the supreme success of the old quantum theory, already contained in germ the principles which would triumph in the new wave and quantum mechanics.

Chapter 8 | Wave Mechanics

1. The Origins and Fundamental Ideas of Wave Mechanics

AROUND 1923 it had become almost obvious that the theory of Bohr and the old quantum theory constituted only an intermediate stage between classical conceptions and some very new conceptions which would permit us to penetrate more deeply into an analysis of quantum phenomena. In the old quantum theory the conditions of quantization were somehow veneered onto the results of classical mechanics. The essentially discontinuous nature of quantization, expressed by the appearance of whole numbers, the quantum numbers, in the formulae presented a strange contrast with the continuous nature of the motions visualized by the old dynamics, either Newtonian or Einsteinian. From all the evidence, we had to succeed in setting up a new mechanics where quantum ideas would come to take their place at the very base of the doctrine and not be added on *ad hoc* as in the old quantum theory. Curious thing! The realization of this program took place almost simultaneously in two very different ways due to the efforts of researchers whose inclinations were basically very dissimilar. Thus there were set up wave mechanics on one hand and quantum mechanics on the other, doctrines whose appearance and formalism at first seemed quite opposite. We shall explain why these theories, so different in appearance, could in reality be considered as identical, each of them being only a mathematical translation of the other into a different language. These two attempts, at the beginning so divergent, to set up a new mechanics truly imbued with quantum conceptions, have finally become blended together into a unique whole which can be called the new quantum theory.

The birth of wave mechanics (1923) is a little prior to that of quantum mechanics (1925). Moreoever, the first lends itself better than the second to an exposition stripped of mathematical formalism. Therefore, for these reasons, wave mechanics will be discussed here first, the following chapter being devoted to quantum mechanics and to a synthesis of the two theories.

First of all we should like to review the reasons which led us, in 1923-24, to state the fundamental ideas of wave mechanics. At that time the discovery of the Compton effect and the study of the photoelectric effect of X-rays had just brought remarkable confirmation to the Einsteinian conception of light quanta. The discontinuous structure of radiation and the existence of photons could scarcely be contested any longer. Henceforth, the formidable dilemma of waves and corpuscles as far as light was concerned was posed with increased sharpness. It was necessary to assume willy-nilly that the picture of waves and the picture of corpuscles had to be used one after the other for a complete description of the properties of radiation, and the relation between frequency and energy that Einstein had put at the base of his theory of photons clearly indicated that this duality of aspect for radiation was intimately connected with the very existence of quanta. From then on, it could legitimately be asked if this strange duality of waves and corpuscles, of which light furnishes so remarkable and disconcerting an example, does not introduce into the scheme of phenomena the profound and hidden nature of the quantum of action and if we ought not to expect to find a duality of the same order everywhere where the constant of Planck manifests its presence. But then a question is asked almost of itself: since the existence of stationary states for atoms shows the intervention of the quantum of action in the properties of the electron, should it not be supposed that the electron presents a duality of aspect analogous to that of light? At first view, such an idea would seem very daring for, up until then, the electron had always proved to be entirely comparable to an electrically charged material point, obeying the laws of classical dynamics (amended in certain cases by the relativity corrections introduced by Einstein). Never had the electron clearly manifested undulatory properties analogous to those of light in the phenomena of interference and diffraction. To attribute wave properties to an electron, in the absence of all experimental proof, could seem like a fantasy of little scientific character. But, nevertheless, as soon as we had the idea that it would perhaps be proper to endow the electron, and more generally material corpuscles, with an undulatory aspect, some troubling statements came to mind. We have explained in Chapter I how the theory of Jacobi had permitted classical dynamics to group the possible trajectories of a material point in a given field in such a fashion that the trajectories of the same group are comparable to the rays of a wave propagation in the sense of geometrical optics. This remarkable parallelism allowed the principle of Least Action to be considered as a kind of translation of the principle of Minimum Time of Fermat. Assuredly this formal identity between a certain manner of representing dynamics and geometrical optics had not escaped those penetrating minds such as the mathematician Hamilton, but it does not seem that they had sought to attribute a physical meaning to it. Moreover, various circumstances seemed to oppose doing so. First and foremost, the theory of Jacobi established a correspondence between the propagation of a wave and a group of *possible* trajectories of the given corpuscle. But in classical conceptions, the corpuscle, in each physically realized case, describes a perfectly well-determined trajectory, and the group of possible trajectories is an abstraction that the mathematician has a perfect right to visualize, but to which the physicist does not seem capable of attaching a concrete significance. In the second place, certain divergences in mathematical form seemed to indicate that the motion of a corpuscle could not really be physically compared to the propagation of a wave: thus, if one wishes to equate the velocity of a corpuscle to the velocity of a wave, we are troubled by the fact that these two velocities do not figure in the same way in the principle of Maupertuis on one hand and in that of Fermat on the other. Despite these well-known difficulties, it was troubling, when the ideas outlined above had been conceived, to note that in classical analytical mechanics the formal analogy between trajectories and the rays of a wave propagation was established through the intermediary of action, that is, precisely through the quantity which serves as the vehicle for quanta. Did this not act, in truth, to confirm the opinion according to which the quantum of action would serve as a bond between the corpuscular and the undulatory aspect of material points?

And then, other indications pointed in the same direction. If it were true that the electron, in macroscopic phenomena, had always been treated as a simple corpuscle, were we not obliged, in order to express its manner of existing in the interior of an atom, to impose on it the strange conditions of quantization where whole numbers appeared? Such a way of restraining the application of classical dynamics in extending it to the electron rather marked its insufficiency and clearly indicated that the properties of the electron were not always those of a simple corpuscle. Upon reflection, this appeal to whole numbers in order to characterize the stationary states of atomic electrons even seemed quite indicative. Whole numbers are frequently met with, in fact, in all branches of physics where waves have to be considered: in elasticity, in acoustics, in optics. They appear in the phenomena of standing waves, of interference, and of resonance. It was therefore admissible to think that the interpretation of the conditions of quantization would lead to introducing a wave aspect of the intra-atomic electrons. To endeavor to attribute to the electron, and more generally to all corpuscles, a dualistic nature analogous to that of a photon, to endow them with a wave and a corpuscular aspect interrelated by the quantum of action, therefore, seemed to be an urgent and productive task.

2. *The Corpuscle and Its Associated Wave*

In the main, what was the problem? Essentially this—to establish a certain manner of associating the propagation of a certain wave with the motion of any corpuscle, the quantities which specify the wave being related to the dynamic quantities by relations which contain the constant h. And it was desirable to establish this association so that the general rules expressing the connection between the wave and the corpuscle would, when applied to the photon, yield the well-known and well-verified relations established by Einstein expressing the association of photons and light waves.

In order to attack the problem as it was thus set up, it was most natural to consider the simplest case: that of a corpuscle in uniform, rectilinear motion with a given, constant energy and momentum. Considerations of symmetry obviously imposed associating it with a wave travelling in the direction of motion. It remained to find out how the frequency and wave length of this wave were connected with the dynamic quantities which characterized the associated corpuscle. Arguments based on the general principles of the theory of relativity then led us to the following result: the frequency of the associated wave is equal to the product of the energy of the corpuscle by Planck's constant, the wave length of the associated wave is equal to the quotient of Planck's constant by the momentum of the corpuscle. This connection between the corpuscle and its associated wave had the great advantage of being exactly the same as that which Einstein had made use of in associating the photon with the light wave. Thus there was realized a notable synthesi since the same duality was shown to be established for corpuscle of matter and for light.

Furthermore, a different way led equally to the same method of establishing the connection between a corpuscle and its associated wave. We have said that Jacobi's theory very clearly suggested the idea of identifying the trajectory of the corpuscle with the ray in a certain wave propagation by identifying the integral of action of the corpuscle with Fermat's wave integral in a way to make the principle of Least Action coincide with the principle of minimum time. By thus proceeding, we immediately find again the proportionality of energy and frequency on one hand and of the momentum and the reciprocal of th wave length on the other; it then suffices to put the constant of proportionality equal to h (which is quite natural and consistent with the idea of uniting the two terms of the duality by means of the quantum of action) in order to reestablish the correspondence already set up by the relativity method. And thi new train of argument does not appeal explicitly to relativit conceptions and, consequently, can be developed within the framework of Newtonian dynamics.

It is also easy to deduce from these original results a most important consequence relative to the connection between the associated wave and the velocity of the corpuscle. In wave theory we must visualize, along with monochromatic waves of given frequency, limited wave groups formed by the superimposition of different monochromatic waves. Among these group it is important to consider those which are formed by the superimposition of monochromatic waves whose frequencies lie within a very small spectral interval about a central frequency. In fact, as we have already had occasion to remark, a monochromatic wave is an abstraction which is never realized in practice: what we call monochromatic waves in experiments are always groups of waves occupying a small spectral interval. Now, if the propagation of a group of waves is examined in those conditions where the velocity of propagation of the monochromatic waves is a function of the frequency, it is discovered that the group of waves possesses a velocity of the whole which is distinct from the velocity of propagation of the waves which constitute the group. This group velocity is expressed as a function of the central frequency of the group and depends on the change of the individual wave velocities as a function of the frequency: it is given by a formula called the "formula of Rayleigh" from the name of the famous English physicist who first pointed it out. We can try to apply this theory of group velocity to the wave associated with a corpuscle as we have just conceived it and we then can set up a correspondence between a corpuscle in uniform rectilinear motion with a given energy and the propagation of a group of waves in the same direction whose frequency is equal to this energy divided by h. This application of Rayleigh's formula then shows that the velocity of this wave group is equal to the velocity that classical mechanics attributes to the corpuscle in question. This remarkable coincidence is very satisfactory, for it means that the corpuscle remains linked with its wave group during the course of the motion. Besides, general wave theory has taught us that the group velocity is nothing else than the velocity of transport of energy by waves. Since, in our dualistic conception, energy is attached to the corpuscle, it is natural that the group velocity of the associated waves should be equal to that of the corpuscle.

These first satisfactory findings were incomplete since they applied only to the very special case of rectilinear uniform motion of a corpuscle in the absence of a field. But it was not very difficult to generalize them. Let us consider, for example, the motion of a corpuscle in a constant field. Jacobi's theory teaches us to regard the trajectory of the corpuscle as the ray of a certain wave propagation and, by the identification of the principle of Least Action and that of Fermat, we come back again to the same relations linking the corpuscle and its wave: the energy (constant) of the corpuscle is equal to the frequency of the wave multiplied by h, the momentum of the corpuscle, varying from one point of the field to the next, is equal to h divided by the wave length of the associated wave, equally variable in space. We can still generalize further by considering fields which vary in time: everywhere we again find relations of the same form between the dynamic quantities of the corpuscle and the quantities of the associated wave such as frequency and wave length.

A noteworthy application permits us to see that by so generalizing the parallelism between the corpuscle and its associated wave, we are going in the right direction. If, indeed, we examine how the waves associated with electrons according to the above theory behave in the interior of the Bohr atom, we arrive at an understanding of the true meaning of the conditions of quantization: they express the fact that the wave associated with the electron is in resonance with the length of its trajectory; in other words, they express the fact that the wave associated with a stationary state of the atomic electron is itself a stationary wave in the sense of the wave theory.

In order to understand the real importance of this result, we must briefly recall what a stationary wave is. When a bounded medium is capable of propagating waves of any kind whatsoever, stationary waves can be set up in this medium, i.e., vibrations whose configuration in space does not change in the course of time. The form of these vibrations is at once determined by the nature of the equation for the wave propagation, by the form of the boundaries of the given medium, and by the conditions which prevail at these boundaries. For example, it frequently happens that the conditions prevailing at the boundary of the medium constrain the vibrations to be zero at these boundaries (vibrating strings fixed at both ends, radio antennae insulated at both ends, etc.): we then must seek solutions of the wave equation which are periodic with respect to the time, and zero at the boundaries of the medium, and whose amplitude is everywhere finite, single-valued and continuous within the medium. This solution constitutes the mathematical problem of the determination of the "proper values" of an equation in derivatives or in partial differentials for a certain region of space and certain boundary conditions. Many simple examples are well known to all physicists; such as the elastic stationary waves to which a vibrating string fixed at both ends can give rise and whose frequencies are multiples of a fundamental frequency; and also the electromagnetic stationary waves which can exist in a radio antenna, insulated at one end and grounded at the other, stationary waves whose wave lengths are equal to four times the length of the antenna divided by successive odd integers.

The considerations of wave mechanics to which we made reference when applied to the atom, lead us to consider that the stationary states of Bohr are those which correspond to stationary waves associated with the atomic electrons. It is undeniable that this interpretation throws a flood of light on the real meaning of the quantum conditions and makes extremely probable the exactness of the fundamental ideas which we have outlined above and of the way in which they lead us to associate waves with corpuscles. Nevertheless, it underscores two difficulties that we should like to point out at this time, because it is most important to study them in order to have a good comprehension of the topics to be discussed in the following sections.

The first difficulty arises from the fact that in order to show the stationary character of the waves associated with a stationary state of the atom, we used formulae which associate the motion of a corpuscle with the propagation of a wave *in the sense of geometrical optics*. By transposing, as a matter of fact, into quantum terms, ideas which are very well known in analytical mechanics, we have set up a correspondence between the trajectories of a corpuscle as conceived in the classical manner, and the rays of a wave propagation. We have already pointed out Chapter 2, sec. 3) that geometrical optics, judged from the general point of view of the wave theory, is only a first approximation, valid when the propagation is freely carried out without meeting any obstacles and when, in addition, the velocity of propagation does not vary too quickly from one point in space to the next. Now it is easy to recognize that the second condition certainly is not fulfilled within an atom for waves associated with an atomic electron, and hence, the way in which we obtained proof of the stationary character of the wave attached to a quantified state of the atom can not be considered rigorous. To put the problem exactly, it would be necessary to set up an equation of propagation for the waves associated with an electron and to solve the problem of the proper values, which then arises, for the waves obeying this equation within our atom. We shall see in the following section how this problem was solved, and has, all in all, led to conclusions in accord with the approximate deductions of the beginning. But we must emphasize a general idea which is contained in the preceding considerations. This important idea is the following: since geometrical optics is only an approximation, valid in certain conditions, and since we have been led to set up a correspondence between classical dynamics and wave propagations, defined as they are in geometrical optics, it would now seem to us that classical dynamics is without doubt only an approximation having the same limits as geometrical optics of which it is in a sense a translation. In all cases where the wave associated with a corpuscle is not propagated according to the laws of geometrical optics (and we have just seen that this is the case for waves associated with electrons in quantized atomic systems), the dynamic evolution of the corpuscle can not be interpreted by means of concepts and laws of classical mechanics. It was therefore necessary to classify henceforth Newton's mechanics and even that of Einstein as being the "old mechanics" and to create a new mechanics in whose framework the old mechanics will appear as first approximations, valid under certain conditions. In brief, it seemed necessary, as we wrote at that time, to set up "a new mechanics with a wave character which would be with respect to the old mechanics what wave optics is with respect to geometrical optics." We shall soon see how this idea was made definitely precise by the memorable works of Schrödinger.

There still remains the examination of the second of the difficulties of which we spoke above. In order to get at its essence, let us consider, as a simple example of a system giving rise to stationary waves, a vibrating string fixed at both ends. This string is capable of giving rise to an indefinite series of stationary waves. The case when the string executes just one of the stationary vibrations, i.e., when it possesses a strictly sine motion, is obviously exceptional; the string after any initial disturbance whatsoever, has, in general, a complicated motion except at the fixed extremities where naturally the motion is always zero. But the mathematical theory of Fourier series tells us that the motion of the string, whatever it may be, can be resolved into a sum of stationary vibrations: this result is expressed by saying that the sine functions representing the stationary waves form a complete system of orthogonal functions. This result may be generalized for vibrating systems less simple than the elastic string fixed at its extremities: it can be shown that, when a region of space is capable of being the seat of stationary vibrations, any vibration whatsoever can be considered as the superimposition of a certain number (finite or infinite) of stationary vibrations. The application of these general ideas to quantized atomic systems at once raises the above-mentioned difficulty. In the original conceptions of Bohr, the atom always had to be in one or the other of its stationary states; the discontinuity implied by quanta being assumed, there was nothing there to contradict the classical picture of a state of the atom. But if it is assumed that the stationary states correspond to stationary vibrations, the general theory that we have just stated leads us to say: the state of an atom at a given instant can be reduced only exceptionally to a single stationary state; in general, it is formed by the superimposition of a certain number of stationary states. With classical conceptions, such a statement has, so to say, no meaning for it can not be imagined that an atom would be in several states at the same time. This difficulty shows us that the development of a new mechanics is going to demand a profound modification of the fundamental concepts of classical physics, a modification whose necessity, as we have said, is already contained in embryo in the very existence of the quantum of action. It is the probability interpretation of the new mechanics which will soon permit us to give meaning to the superimposition of several states.

3. *The Work of Schrödinger*

To Erwin Schrödinger came the distinction of being the first, in his magnificent papers that appeared in 1926, to write out explicitly the wave equation of wave mechanics and to have deduced from it a rigorous method of studying the problems of quantization. To write the equation of the waves associated with a corpuscle in wave mechanics, we can start with this idea that in the eyes of the new theory the old mechanics is equivalent to the approximation of geometrical optics. In the theory of Jacobi, the trajectories of a corpuscle are regarded as being identical to the rays of a wave propagation whose wave surfaces are defined by a complete integral of an equation in partial derivatives of first order and second degree, called "Jacobi's equation." We have already remarked [1] that Jacobi's equation has a form quite analogous to the fundamental equation of geometrical

[1] See Chap. 2, sect. 2.

optics and that is precisely the reason why there is an analogy between the theory of Jacobi and the theory of wave propagation in its geometrical approximation. The wave equation of wave mechanics therefore must be chosen in such a way that the corresponding equation of geometrical optics, an equation valid under the conditions which we have already made precise, would coincide with the equation of Jacobi. To set up an equation of propagation satisfying this condition, Schrödinger outlined the following procedure: first, an expression is set up which, for the given problem, would give in classical mechanics the value of the energy as a function of the coordinates of the corpuscle and of the components of its momentum; then, in this expression, which in mechanics is called the Hamiltonian, each of the rectangular components of the momentum is replaced by the symbol for derivation with respect to the coordinate multiplied by a constant proportional to the constant h of Planck. Thus the Hamiltonian has been transformed into a symbol of operation, the Hamiltonian operator. It is then sufficient to apply this operator to the wave function of the system (which is always designated by the Greek letter ψ) and to put the result obtained equal to the derivative of the wave function with respect to time multiplied by the constant which has just been mentioned; the equation thus obtained can be taken as the wave equation of the corpuscle for, in the approximation of geometrical optics, it reduces to the equation of Jacobi such as it would be written in classical mechanics for the problem at hand.

We must make a few remarks concerning the propagation equation thus obtained for the wave associated with a corpuscle. First, this equation defines the wave function as a scalar function and not as a vector. This establishes an important difference between the wave associated with a corpuscle and with a light wave. But it is known that the wave theory of light, too, had started by considering light as being defined by a scalar quantity, the light variable, and this point of view can be held today in interpreting many of the phenomena of diffraction and interference. It is only when we wish to account for polarization that it is necessary to introduce the vectorial character of the wave function. So might one think that the scalar wave function ψ would one day be replaced by a wave function of several components through extension of the theory: later on this prediction will prove to be confirmed in the birth of the theory of the magnetic electron due to Dirac, without however this resulting, as we shall see, in a complete similarity between the theory of the electron and that of the photon.

The second remark to be made about the equation of propagation is that it is complex, i.e., its coefficients are not all real numbers and the quantity $\sqrt{-1}$ figures in it. This circumstance, at first glance rather odd, shows us how difficult it is to give the ψ-wave of wave mechanics the same physical significance that the waves pictured by classical physics were considered to possess. In classical physics, indeed, the quantities which are propagated by waves are derived from vibrations of a medium whose existence is certain or supposed (this latter is the case of the ether in the classical theory of light): therefore they must, since they represent an actual phenomenon, be expressed by a real function. If, sometimes, it is judged useful, as it often happens in optical calculations, to replace these real numbers by the complex quantities of which they are the real parts, that is only an artifice of calculation and can always be dispensed with. In wave mechanics, on the contrary, by reason of the presence of imaginary coefficients in the equation of propagation itself, the complex character of the ψ wave function is shown to be essential and resists all attempts to consider the wave of wave mechanics as a physical reality corresponding to vibrations in any medium. The development of the new mechanics has led to a consideration of the quantity ψ as a simple intermediate quantity, the knowledge of which permits us to calculate certain other quantities, these latter real, having a physical significance, more of a statistical kind. We shall have to return to this point, but it was pertinent to note for the moment why the equation of propagation in wave mechanics obliges us, by its very form, to give up a physical interpretation of the associated wave.

We have just explained how Schrödinger had succeeded in setting up the equation of propagation of the ψ-wave associated with a corpuscle for the general case. But he started, in order to find it, from the formulae of Newtonian mechanics, so that this equation of propagation does not satisfy the demands of the theory of relativity. It is therefore natural to think that this equation is valid only for corpuscles of quite low velocities, i.e., for waves of not too high frequency, and the question is asked to find an equation of propagation having a relativistic character and containing that of Schrödinger as a first approximation for low frequencies. Several writers proposed, almost simultaneously, an equation of this kind, which presents itself naturally to the mind. But this relativistic equation of propagation, which is of second order with respect to time, has led to difficulties and we shall see that the true relativistic generalization of the original equation of propagation was obtained by Dirac in another way.

Schrödinger has also given the form of the equation of propagation (non-relativistic) which is appropriate for a system of corpuscles, i.e., for an assemblage of corpuscles interacting on each other. But since new conceptions, requiring a special study, are introduced, we shall put off until a later chapter (Chapter 12) an examination of the wave mechanics of systems of corpuscles.

Armed with his equation of propagation, Schrödinger was able to attack rigorously the problem of the determination of the stationary states of a quantized system by assuming, according to the indications of the approximate theory, that these stationary states correspond to the stationary forms of the associated wave. Let us consider a quantized system such as the hydrogen atom. We know the equation of propagation of the associated wave in this system and it is natural to assume, since the system is concentrated within a region of space, that the ψ-function rapidly tends to zero as we go away from the center of the system. If we also assume, as is usual in mathematical physics, that the ψ-function must everywhere be single-valued and continuous, the determination of the stationary waves will be made by seeking monochromatic solutions of the equation of propagation which are finite and single-valued in the whole of space and which are zero at infinity. Schrödinger solved this problem brilliantly, by making use of known methods of analysis, for several simple types of quantized systems. It was discovered that there exist monochromatic solutions satisfying the imposed conditions only for certain particular values of the frequency; these values are the "proper values"[1] of the propagation equation in partial derivatives for the given problem and with the boundary condition that ψ is zero at infinity. To these proper frequencies of the given system correspond, through multiplication by h in agreement with the general relation between wave and corpuscle, the quantized values of the energy of the corpuscle. The calculations of Schrödinger should therefore furnish for the cases studied the quantized energies and hence the spectral terms. In a great number of cases, exactly the same result is found as in the old quantum theory: this happens, for example, in the case of the hydrogen atom where the results of Bohr are exactly recovered. But in other important cases the results are different from those which the old quantum theory furnished and the new results are in better accord than the old with the indications of experiment. The most remarkable example of this is that of the linear oscillator. It will be recalled that the quantization of the linear oscillator, encountered by Planck in his radiation theory, had been the point of departure for the whole development of the quantum theory. The old method of quantization assumed that the quantized values of the energy of the linear oscillator were whole multiples

[1] Throughout this book, "proper value" and "proper function" have been used in place of the more usual, but less expressive and comprehensible terms "Eigenvalue" and "Eigenfunction." Tr.

of a quantum of energy obtained by multiplying the proper frequency of mechanical oscillation of the linear oscillator by h. Now certain physical phenomena in the theory of which the quantization of the oscillator appears (band spectra of bi-atomic molecules, for example) seem to indicate that the quantized energies of an oscillator are equal, not to the product of its quantum of energy by an integer, but to the product of this quantum by a half-integer, i.e., by a number of the series $1/2$, $3/2$, $5/2$, . . . $(2n+1)/2$ . . . Now, the new method of quantization, departing here from the old quantum theory, predicted precisely this quantization by half-integers. In this way Schrödinger found again the correct results of the old theory and corrected others: his success was complete.

Then it was that a curious coincidence struck Schrödinger and put him on the road toward one of his most fortunate results. Heisenberg's quantum mechanics had been developed shortly before this time. Now this new method, quite distinct in appearance from wave mechanics, gave exactly the same results for the value of the quantized energies of atomic systems as Schrödinger's method did, confirming or correcting in a like manner the results of the old quantum theory. Schrödinger had an intuition that this coincidence could not be due to chance and he was skillful enough to show that quantum mechanics, despite its absolutely different aspect, is only a mathematical translation of wave mechanics. We shall limit ourselves now to pointing out this fine work of Schrödinger, since we intend to return to it in the following chapter.

The importance of the Zeeman effect and its electric analog, the Stark effect, is well known. Schrödinger wanted to attack the theory of these phenomena by wave mechanics. In order to do so, he developed a good method of perturbations which is a wave translation of the classical method in celestial mechanics. Magnetic or electric fields that we can produce are, indeed, very feeble compared to the fields that are acting within atomic systems. Therefore when we subject atoms to either a uniform magnetic or electric field in order to produce the Zeeman or Stark effects, this field can be considered as a very small perturbation of the field which normally acts within the atomic system. If we have already succeeded in calculating the quantized values of the energy for a given system in the absence of a field, it will be necessary to calculate the very slight modification that the presence of a perturbing field would impose on these quantized values. This problem was solved by Schrödinger through his method of perturbations and led him to a detailed prediction of the Zeeman and Stark effects. For the Stark effect the results are on the whole in accord with those of the old quantum theory, but would seem to be more exact in several respects. For the Zeeman effect, we find again, in accord with the old quantum theory, the classical predictions of Lorentz. This is satisfactory in this case since, in the main, these things actually do occur as Lorentz had predicted (normal Zeeman effect). But beyond the normal Zeeman effect satisfying the predictions of Lorentz, in many cases there are also quite complicated anomalous effects; neither classical theory, nor the old quantum theory, were able to explain these complex facts. Wave mechanics in the hands of Schrödinger was no more successful at it. In order to interpret the anomalies of the Zeeman effect, a new element had to be introduced: the spin of the electron. We shall speak about this matter in a later chapter.

We shall also put off until the following chapter the study of that part of Schrödinger's work relative to the emission and dispersion of light.

4. The Diffraction of Electrons

We have just recounted how the ideas of this author on the relationship between waves and corpuscles and the necessity of constructing a new mechanics with a wave character had, thanks to the admirable papers of Schrödinger, taken on by 1926 an extraordinary fullness and precision. But however nice the general ideas and fundamental methods might be, however precise the verifications which they had received from a correct prediction of atomic phenomena might seem, there was still lacking a direct, experimental verification for these conceptions. The year 1927 brought about this verification through the discovery of the phenomenon of electron diffraction by Davisson and Germer.

Since the motion of corpuscles is intimately connected with the propagation of a wave, one might wonder if material corpuscles, electrons for example, might not exhibit phenomena of interference and diffraction quite analogous to those which photons exhibit and whose study makes up physical optics. In order to see which of these phenomena are those which might actually be observed, it was pertinent to evaluate first of all the wave length of the wave associated with the electrons that we know how to use in usual cases. The formulae of wave mechanics immediately furnished a precise answer to this question: the wave length associated with the electrons in usual circumstances is always very small, of the order of that of X-rays. What we might therefore hope to obtain with them are those phenomena that can be obtained with X-rays. Now, it is known that the fundamental phenomenon of X-ray physics is the phenomenon of their diffraction by crystals. The extreme smallness of the X-ray wave length makes it almost impossible to use apparatus made by the hands of man to obtain their diffraction. Fortunately, nature has given us gratings adapted to this diffraction, namely, crystals. In crystals, as a matter of fact, the atoms or molecules are regularly distributed and form a three-dimensional grating, and it is found that the distance between the material centers distributed throughout the crystal are always of the order of magnitude of the length of the X-rays. By sending a beam of X-rays into a crystal, the phenomenon of diffraction should therefore be obtained analogous to that which is obtained with light when a 3-dimensional point grating is used. It is known that the phenomenon of the diffraction of

X-rays by crystals was effectively brought to light in 1912 by von Laue, Friedrich and Knipping and that today it serves as the basis of the considerable development in X-ray spectroscopy. According to what we have just said, we might expect to be able to obtain a quite analogous phenomenon with electrons. By using a beam of electrons with known kinetic energy, we should be able to observe a diffraction phenomenon quite analogous to those which can be produced with X-rays. The structure of the various crystals used in this kind of experiment being well known today through various methods, principally through Röntgen spectra, we should be able to deduce from the diffraction figures so obtained the wave length of the wave associated with the electrons that were used and consequently, we would be able to verify the exactness of the relation as proposed by wave mechanics connecting the wave length of the associated wave with the motion of the corpuscle.

It was to Davisson and Germer, working at the Bell Telephone Laboratories in New York, that fell the honor of discovering the diffraction of electrons by crystals. By bombarding a crystal of nickel with a beam of monokinetic electrons, they clearly detected that the electrons were diffracting as a wave of a given wave length should do, and they showed that this wave length is just that which the formula of wave mechanics predicts. So the existence of this fine phenomenon was established, a phenomenon whose simple statement a few years previously would have provoked the astonishment and aroused the incredulity of physicists.

Repeated almost at once in England by G. P. Thomson, son of Sir J. J. Thomson, using a somewhat different method, the diffraction of electrons was soon repeated almost everywhere. By varying the conditions and experimental arrangements, Ponte in France, Rupp in Germany, Kikuchi in Japan, and many others, studied this phenomenon which was soon known in all its details. Most of the small difficulties of interpretation which were present at first were soon cleared up, principally by noting that the interior of a crystal has an index of refraction different from unity for waves associated with electrons. As had previously been done for X-rays (Compton, Thibaud) the diffraction of electrons was successfully obtained with a simple, ordinary grating by using an almost tangential incidence (Rupp). In this way the electron wave length can be directly compared with lines traced on a metallic surface by a mechanical device.[1]

As often happens, the phenomenon of electron diffraction, which at first seemed most difficult to obtain and had demanded a very great skill on the part of the experimenters who succeeded in detecting it, has since become a relative easy and everyday occurrence. The technical arrangements to bring it about are so thoroughly perfected that today electron diffraction can be demonstrated to students in a lecture hall! Then again, the conditions of these experiments have been varied over so wide a range that the correctness of the fundamental formulae, expressing the relation between wave and corpuscle can now be asserted to hold throughout an enormous energy interval which extends from a few score electron-volts to a million electron-volts. For large values of the energy, it naturally is necessary, when verifying the formulae, to keep the terms which introduce relativity corrections. Relativistic conceptions therefore receive through this an indirect confirmation.

The validity of the formulae which give the length of the wave associated with a corpuscle has become so certain that today electron diffraction is used not to verify these, but, by assuming them, to study the structures of certain crystallized or partially oriented media. But these are somewhat technical matters which do not belong in this work. Let it suffice that we point out that electron diffraction experiments have brought a

[1] In 1940, Börsch was able to obtain electron diffraction at the edge of a screen, a phenomenon analagous to one which has been known for light since the time of Fresnel. (Note added in 1946.)

magnificent direct confirmation to the conceptions of an association between waves and corpuscles which have served as the point of departure of the new mechanics.

It will be pertinent to point out, before finishing this section, that diffraction of material particles other than electrons has also been obtained. Protons and material atoms diffract as electrons do. Experiments on this subject are difficult and still not numerous, but it is certain that the formulae of wave mechanics are verified even here. This should be not at all surprising. The association of waves and corpuscles would seem to be a great law of nature, this duality of aspect being connected with the existence and significance of the quantum of action. There is no reason to limit it to electrons and it is not surprising to meet up with it for all physical entities.

5. *Physical Interpretation of Wave Mechanics*

WE MUST now try to find out what use can be made of our knowledge about the wave function of a system. The old mechanics, corresponding to the approximation of geometrical optics—all the images and conceptions that were used in it—all these must be abandoned when we pass beyond the limits of this approximation. We can not therefore avail ourselves, at least without precautions, of the notions of a position, of a velocity, and of a trajectory of a corpuscle. We must take up this matter again and investigate what kind of predictions our knowledge of the wave function will permit as to the observable phenomena relative to corpuscles. The system of postulates which we will elaborate will have to satisfy the essential condition of leading us back to the conceptions and results of the old mechanics whenever the ψ-wave obeys the laws of geometrical optics. As we are going to see, the interpretation of the new mechanics is of a probabilistic nature: however, we are going to undertake an overall consideration of the probability interpretation of the new mechanics in Chapter 10. For the moment, we are going to take up the surface aspects of the question by showing only what postulates the physicists were almost immediately obliged to assume in order to use the equations of wave mechanics.

First of all, since the ψ-function is essentially complex, it can not be regarded, as we have said, as representing a physical vibration, but we can try to form by means of ψ some real expressions having a physical significance. One such which most naturally comes to mind, is the square of the modulus of the complex quantity ψ, a square which is obtained by multiplying the wave function by the conjugate complex quantity. This quantity can be considered as the square of the amplitude of the ψ-wave, i.e., as its intensity in the ordinary meaning of the wave theory. To comprehend what significance we must attribute to this important quantity, we must go back to light theory, which so often has served us as a guide, and find out what the intensity of a light wave represents when the existence of photons is assumed. Let us consider any one of the diffraction or interference experiments which are classical ones in optics. The wave theory determines (and we know with what exact precision) the position of the bright and dark fringes by calculating the intensity of the light wave at each point and by assuming that light energy is distributed in space proportionally to the wave intensity. This hypothesis, which is justified by various arguments in the different elastic or electromagnetic theories of light, can also be considered as a postulate: the principle of interference.

Let us now introduce the photon concept. A beam of light would seem to us to be a stream of photons and an interference or diffraction experiment becomes in our eyes an experiment where, as a consequence of the apparatus used, the photons are distributed in a non-uniform manner in space, being concentrated in the bright fringes and avoiding the dark fringes. Since the predictions of the wave theory are verified very exactly, we must say that the wave intensity calculated by this theory is at each point proportional to the density of photons.

But we have already pointed out (Chapter 5, sect. 4) those curious experiments which show the possibility of obtaining interference with extremely feeble beams of light. In these experiments, interference is produced even when the photons arrive one after the other at the interference apparatus. It must therefore be assumed, in order to explain the eventual appearance, after long exposure, of the usual interference patterns, that the intensity of the wave associated with each photon represents at each point the probability of the photon's being at that point. We are led in this way to pass from a statistical point of view to a probability point of view, and the principle of interference seems to us to be a principle governing the probabilities of the location of photons. But, if we now return to the theory of matter, we see that a wholly analogous principle is going to be imposed on us, for electron diffraction at a crystal takes place exactly as it would take place for photons of the same wave length. Therefore it is once more the intensity of the wave associated with the electrons which determines the probability of their localization in space. So we come to stating the following principle: "the square of the modulus of the ψ-function measures the probability at each point and at each instant that the associated corpuscle will be observed at this point at that instant." We must not hide from ourselves how much a modification such a postulate entails in our conceptions. Since the ψ-wave in general occupies a definite region in space, the corpuscle can be found anywhere in this region. At a given instant, a definite position can not be assigned to the corpuscle, but it can be said only that there is such and such a probability of finding it here or there. And with the notion of a well-defined position, the notions of velocity and trajectory disappear, or at least grow hazy. Everywhere the certainty of the old mechanics gives way to probability. We are glimpsing here an important change in the method employed by science in the representation and prediction of phenomena, a change which embraces important philosophic consequences.

Reserving the study of these questions until later, we shall state here a second principle which was forced on physicists in their interpretation of wave mechanics. This second principle was first formulated, we believe, by Born at the beginning of his excellent studies on the problems of collision in wave mechanics: it can be given the name of the "principle of spectral decomposition." In order to understand the nature of this new postulate, let us consider first the simple case of a corpuscle which moves in the absence of a field. If the wave associated with this corpuscle is a plane monochromatic wave, we know that the energy of the corpuscle has a well-defined value, equal to the product of the frequency of the wave by h. But, from a wave point of view, nothing obliges us to suppose that the ψ-wave is monochromatic: it can just as well be formed, without ceasing to satisfy the equation of propagation which is linear, by the superimposition of plane monochromatic waves and form a wave group. In this case, what will the energy of the associated corpuscle be? This question is embarrassing since many different frequencies figure in the composition of the ψ-wave. Born proposed to resolve this by making another appeal to probabilities. The corpuscle, according to him, does not have a determinate energy, it can have one of the energies corresponding to one of the frequencies represented in the ψ-wave. More precisely this means that, if the energy of the corpuscle is determined, one of these values will be found, without our being able to say *a priori* which one. But what can be said *a priori*, thanks to the new principle introduced by Born, is the probability of obtaining one or the other of the possible values of the energy. In effect, to say that the wave associated with a corpuscle is formed by a superimposition of plane monochromatic waves, means that the ψ-function is expressed mathematically as a sum of terms representing monochromatic waves: each of these terms is qualified by a coefficient that can be called the partial amplitude of this monochromatic component of the spectral decomposition of the ψ-wave and the square of the modulus of this amplitude will be the corresponding partial intensity. The principle stated by Born then consists in asserting that the probability that a measurement of the energy of a corpuscle would furnish a certain value, corresponding to one of the monochromatic components of the ψ-wave, is given by the corresponding partial intensity in the spectral decomposition of the wave. This principle is again quite in accord with what optics might suggest to us.

For, indeed, if a complex light wave falls on a prism or on a grating, the different monochromatic components of the wave are found to be separated after passing through the apparatus, and obviously we must say that the probability that a photon of the single initial beam would go into such and such a separated beam at the end of the operation, is proportional to the intensity of the corresponding monochromatic component in the spectral development of the incident wave. Furthermore, we must consider this question from a more general point of view. Applied to quantized atomic systems, the principle of spectral decomposition gives us the key to a difficulty of which we have already spoken. In a quantized atom, there exists a series of frequencies corresponding to the stationary states of the quantized energies. But for such a system, just as for a vibrating string, it can very well be imagined how a certain state is formed by the superimposition of stationary states, for, by taking as the ψ-function a sum of proper vibrations, a solution of the propagation equation is again obtained because of the linear character of this equation. But in the state represented by this ψ-function, it can no longer be said that the atom is in one of its stationary states: it is somehow in several stationary states at the same time, which is obviously incomprehensible in classical conceptions. With the principle of spectral decomposition, the difficulty is resolved in an unexpected way: the atom in the state we imagined can have any one of the quantized values of the energy represented in the spectral development of its ψ-wave and that with probabilities proportional to the intensity of the corresponding spectral components. Here again, this means that an experiment permitting us to attribute a certain energy to the atom will lead to a value of the energy represented in the spectral decomposition. The probability character of these interpretations gives us a new foretaste of the absolutely new orientation that physical theory is going to have to take.

A comparison of the two principles that we have just stated leads to the Uncertainty Relations, to which the name of Heisenberg is associated. But the study of this important question will be more in place in the general chapter that we wish to devote to the probability interpretation of the new mechanics and we shall take leave of it for the moment.

6. The Theory of Gamow

WE SHOULD like to mention here a most remarkable application of wave mechanics which is due to Gamow. Interest in this application, outside its heuristic value in the domain of radioactivity, lies in showing how the aspects of certain problems are modified when we pass from the old to the new mechanics.

Let us consider a corpuscle on which a field of force acts so as to check its motion. It may happen that the field of force, which we suppose static, vanishes at a certain point and then changes sign. Then the potential function from which it is derived first increases, passes a maximum, then decreases. We say, using figurative language, that we are dealing with a mountain of potential. Will a corpuscle that ascends the slope of this mountain succeed in crossing over to the other side? To this question, classical mechanics made the following answer: yes, if the corpuscle has sufficient energy to attain the summit of the mountain and descend on the other side, thus realizing a crossing of the mountain; but if the corpuscle does not have sufficient energy to attain the summit of the mountain, it can never cross it, for, having exhausted its energy, it will stop on the incline and finally will fall back again.

In wave mechanics, things take place in an entirely different manner. We must now picture the wave propagation associated with the corpuscle. For this wave, the mountain of potential will be, as can be proven, the analog of a refracting medium as long as the potential is less than the available energy of the corpuscle: if the energy of the corpuscle is greater than the summit of the mountain of potential, it will thereupon easily pass over to the other side. Up to this point, there is no difference from the old theory. But if the energy of the corpuscle is lower than that of the summit of the mountain, all that part of the mountain which corresponds to an energy greater than that of the corpuscle, acts as an extinguishing medium for the associated wave. Now, wave theory tells us that, when a wave falls on an extinguishing medium, it penetrates slightly into this medium in the form of a very rapidly dampened wave, of such a kind that if the extinguishing medium has a sufficiently small thickness, a fraction of the wave, in truth, generally very small, can filter through the extinguishing body. This fact has been thoroughly verified in optics. If we carry over these results into our problem in wave mechanics, we see that a sufficiently narrow mountain of potential can be crossed even by a corpuscle whose energy is too small to permit it to attain the summit of the mountain. More precisely, the corpuscle which runs up against the mountain of potential with too small an energy to overcome its summit, has however a certain probability, surely in general very small, but not zero, of finding itself on the other side: this results from the probability interpretation of the associated wave and from the principle of interference. This phenomenon, so characteristic of wave mechanics, is often designated by the picturesque name of the "tunnel effect."

Let us now suppose that a corpuscle is enclosed in a region bounded on all sides by mountains of potential higher than it can surmount. According to classical mechanics, the corpuscle will be a prisoner forever in this valley of potential. For wave mechanics, on the contrary, the corpuscle has a certain, very slight chance of escaping from the valley: it has a certain probability of escape in a unit of time that the formulae of the new mechanics permit us to calculate.

And now, we come to the application of this that Gamow (and at almost the same time as he, Condon and Gurney) has made of the above considerations to the problem of the disintegration of radioactive substances. It is known that a great number of radioactive substances undergo transmutations with the emission of α-rays. It can be supposed that the α-rays pre-exist in the nucleus of the transmutable atoms and are found there as if in a valley surrounded by a mountain of potential. Since Coulomb's law has been verified in the neighborhood of the nucleus up to a very short distance from the latter, the form of the exterior slope of the mountain of potential is known. It is most probable that Coulomb's law finally ceases to be exact at a certain distance from the nucleus: the potential must pass through a maximum and finally decrease, but the interior slope of the mountain is quite unknown. But one fact had greatly surprised the physicists: the α-rays which leave these transmuting nuclei seem to have an energy too low to permit them to surmount the mountain of potential which guards the nucleus. As a matter of fact, the exterior slope of the mountain can be explored far enough to let us know that the summit certainly rises beyond a certain height. Now, the α-rays which leave the nucleus do not have sufficient energy so that they could have reached this summit. With classical ideas, we reach an impasse. But the tunnel effect comes along to explain all. The α-particle enclosed in the nucleus of a transmutable substance is in a valley of potential bounded by mountains whose summit it can not attain: nevertheless, it has a certain probability in unit time to escape to the exterior and this probability obviously is equal to the disintegration constant of the radioactive substance. Wave mechanics would therefore permit us, if we knew exactly the form of the mountain of potential bounding the nucleus, to calculate the disintegration constants of radioactive substances by these

α-rays. By making some plausible hypotheses on the form of these mountains, Gamow has shown that results very close to reality are reached.

One of the principal successes of Gamow's theory is having explained the Geiger-Nuttall law according to which the emission velocity of the α-rays is greater for elements of short half-life than for elements of long half-life. This law is expressed mathematically by a relation between the constant of distintegration and the energy of the α-particles emitted by the transmutation, a relation from which it follows that the constant of disintegration varies rapidly as a function of the energy of the α-particles. Gamow has shown that his theory accounts for this law very closely. The reason for this agreement is easy to understand: the more a corpuscle imprisoned in the valley lacks the energy necessary to attain the summit of the mountain which surrounds it, the smaller is the probability of its escape. And this probability diminishes very rapidly with the energy of the imprisoned corpuscle. Since this probability is equal to the constant of disintegration and since the corpuscle, having escaped by means of the tunnel effect, possesses the same energy as before its escape, we thus find a relation between the constant of disintegration and the energy of the α-particles emitted during transmutation. The form of the law is that which experiment indicates and some plausible hypotheses about the profile of the nuclear mountain of potential permits us to find a numerical agreement.

Gamow's theory is certainly very sketchy, for the nucleus of the heavy radioactive elements is surely something a little more complicated and can not be reduced to a simple valley of potential containing α-particles. Nevertheless, the success of Gamow's theory in the interpretation of certain facts shows the value of the new conceptions of wave mechanics and the necessity of introducing probability considerations in order to resolve certain undeniable difficulties raised by the experimental facts themselves.

Chapter 9 Heisenberg's Quantum Mechanics

1. *The Guiding Ideas of Heisenberg*

HEISENBERG'S INITIAL PAPER on quantum mechanics appeared in 1925, that is, at a date intermediate between the appearance of the first ideas concerning wave mechanics and the publication of Schrödinger's papers. But Heisenberg's aim appeared to be quite different from that which was pursued by these others. The ideas which guided Heisenberg had in fact no apparent relation with those from which the first advances in wave mechanics had sprung, and the formalism on which it was built had quite a special aspect. Let us begin by examining Heisenberg's guiding ideas.

Heisenberg, as we have seen, belonged to that "Copenhagen school" which had grown up around Bohr and he had devoted his first endeavors to applications of the method of correspondence. It was therefore most natural that the spirit of this method, so original and so profound, should impregnate his thoughts. Now one of the essential ideas which emerged from the study of the Correspondence Principle is the following: while classical theory expresses the quantities attached to a quantized system in the form of a Fourier series development where each term corresponds to continuous and simultaneous emissions of radiation, quantum theory resolves these same quantities into elements corresponding to the various quantum transitions of which the atom is capable, each of these elements being connected with discontinuous and individual acts of emission of radiation. As we have explained previously, the goal of Bohr's famous principle was to establish a "correspondence," at least asymptotic, between these two so dissimilar means of representation. What seems to have struck Heisenberg is that in passing from the classical point of view to the quantum point of view it is necessary to break up all the physical quantities and to reduce them to a powder of distinct elements corresponding to the various possible transitions of the quantized atom. Hence, the idea, at first so disconcerting, of representing each physical quantity attached to a system by a table of numbers analogous to what mathematicians call a matrix. Somehow, the Fourier series in the classical representation is pulverized into an infinity of discrete elements, the assemblage of these elements continuing to represent the quantity being considered. Of course, these elements will have to be subject to some such rules so that for large quantum numbers, we reach an asymptotic coincidence by setting up the correspondence between the various transitions and the components of the classical Fourier series as shown by Bohr.

Heisenberg also saw another advantage in adopting this new representation of quantities by an assemblage of matrix elements: he thought in this way to bring about the elimination of unobservable quantities with which previous quantum theories had been encumbered. To use a rather formidable expression from the language of philosophy, he adopted a strictly "phenomenological" attitude and would have liked to eliminate from physical theory all that does not correspond to observable phenomena. What good is it to introduce the position, velocity or trajectory of atomic electrons into our atomic theories since these elements are not capable of observation or measurement? What we know about the atom are its stationary states, its transitions between stationary states and the radiations which are connected with these transitions. We must introduce into our calculations only those elements related to these observable realities. This was the program which Heisenberg wanted to accomplish. In his matrices, the elements are arranged in rows and columns, each one of them being specified by two indices which give the number of the row and the number of the column. The diagonal elements, i.e., those whose indices are equal, correspond to a stationary state; the non-diagonal elements, whose two indices are different, correspond to the transition between the stationary states defined by these indices. As to the value itself of the elements, it will be related, by formulae inspired by the Correspondence Principle, to quantities characterizing the radiations emitted during the transitions. Thus there will be realized a representation where everything will correspond to observable phenomena.

Obviously, one might wonder if Heisenberg actually succeeded in eliminating all that is not observable. The presence of matrices representing the coordinates and the momenta of the atomic electrons in the formalism of his quantum mechanics might leave some doubt in this respect. But the determined attempt of Heisenberg, even if it does not completely fulfill the philosophic program of its author, has resulted in setting up a new mechanics, of most curious aspect, and has led to some remarkable results and constitutes an essential stage in the development of the new quantum theories.

2. *Quantum Mechanics*

IT IS most difficult to outline quantum mechanics, even quite superficially, without making use of a mathematical formalism, for it might be said that the essence of this new mechanics is precisely its formalism. Nevertheless, we are going to try to give here a vague idea of what quantum mechanics is, this mechanics of matrices whose origin is due to Heisenberg and whose development is due both to him and to Born and Jordan.

Heisenberg, then, started out with the idea of substituting tables of numbers, matrices, for the physical quantities that are usually dealt with in atomic theory. Guided by the method of correspondence, he first sought to establish rules for the addition and multiplication of these various matrices, each of which is considered as a single mathematical entity. He found that these rules for addition and multiplication are exactly the same as those for the matrices that mathematicians had been used to using in the theory of algebraic equations or in that of linear substitutions. This result, which is not at all evident *a priori*, greatly simplified matters, for the properties of algebraic matrices had been very well known for a long time. A singular property of matrices is that multiplication is not commutative for them: it depends on the order of the factors. The product of one matrix by a second does not equal the product of the second by the first. Therefore Heisenberg was representing physical quantities by numbers which do not possess the property of commutative multiplication. This fact might be considered as the very basis of quantum mechanics and Dirac in his first work took this point of view: for him, the transition from classical physics to quantum physics is made "simply" by representing physical quantities, no longer by ordinary numbers, but by "quantum numbers" for which multiplication is not commutative. A great many physicists at that time found that this change was not so simple. Heisenberg must find some way to bring the quantum of action into his theory. Here again, he started out from the way in which the constant h is brought into classical equations by the old quantum theory and he sought to carry over by means of correspondence this introduction of h into his new mechanics. This result is very precise but a little surprising at first. It must be supposed that, in the product of the matrix corresponding to a coordinate by the matrix corresponding to the conjugate component of momentum, the order of the factors is not immaterial and that the difference between the product obtained by taking the two factors in a certain order and the product obtained by taking the factors in the reverse order is equal to Planck's constant multiplied by a numerical constant. All the other canonical variables of quantum mechanics commute, i.e., their product does not depend on the order of the factors. It is only when the product of two quantities, which are canonically conjugate in the sense of analytical mechanics, is considered, that their is a lack of commutation as measured by the quantity h. In macroscopic phenomena where h is negligible, all mechanical quantities can be regarded as commutative and we return once more to classical mechanics, as we must. This way of introducing Planck's constant through relations of non-commutation, although natural from Heisenberg's point of view, may seem a little singular. Later on we shall see the interpretation of it in wave mechanics.

Having thus made precise the properties of the matrices that he was using to represent physical quantities, Heisenberg had to develop the equations representing the evolution of these matrices in the course of time: in other words, he had to set up his dynamics. This he did by boldly assuming that the matrices obey equations identical in form to those of classical mechanics. According to this hypothesis, the canonical equations of Hamilton can be written for the matrices. This identity of the dynamic equations is, however, more apparent than real, for, in classical dynamics, the quantities figuring in the equations are ordinary numbers, while in the mechanics of Heisenberg, they are matrices: important differences arise from this fact. Be that as it may, it can be shown that the canonical equations of quantum mechanics allow us to recover the principle of the conservation of energy, and they are in accord with Bohr's law of frequencies. Furthermore, for atomic systems, these equations, for reasons which we can not develop here, can be satisfied only for certain particular values of the energy. So we find again the existence of stationary states with quantized energy and we are in possession of a method of calculation for these energies. Applying this method at once to the more classical types of quantized systems, Heisenberg and his disciples calculated the quantized energies of the linear oscillator, the hydrogen atom, etc. They found results often in accord with the old quantum theory, but sometimes quite different. For the linear oscillator, for example, they obtained in place of the law of whole quanta assumed by Planck, the law of half-quanta which we have already mentioned as being in better accord with the facts.

Fired by the very interesting results of quantum mechanics, by the rigor and precision of its formalism, a crowd of theorists followed in the steps of Heisenberg, bringing him many new, important contributions. Schrödinger published his papers and noted with astonishment that the method of quantization of wave mechanics led to the same results as the method of quantum mechanics, so different in inspiration. He had the intuition that this fact might not be purely fortuitous, and succeeded in explaining it in an excellent paper which we shall now analyze.

3. *The Identity of Quantum Mechanics and Wave Mechanics*

THE IDEA which directed Schrödinger in his work was that it must be possible to construct, by means of the wave functions of wave mechanics, quantities having the properties of the matrices of quantum mechanics. Quantum mechanics will then seem to be a method permitting us to calculate these quantities and to operate on them without passing explicitly through the intermediary of the wave function. The identity of the two forms of the new mechanics will then be proved.

Wave mechanics, when it treats a problem of quantization, determines the various stationary waves of the system under consideration, and calculates the corresponding wave functions. These functions are called the "proper functions" of the system: they form a sequence which we here shall suppose to be discontinuous since this actually happens in many of the important cases. Let us suppose that we imagine all the ways of pairing all these proper functions, two by two. We shall thus obtain two kinds of pairs: those which correspond to the combination of a proper function with itself and those which correspond to the combination of a proper function with a different proper function. The first are attached to a single stationary state: the others are attached to two different stationary states and can consequently be considered as being attached to the transition between these two stationary states. We obtain then, by these combinations of proper functions two by two, a series of elements that can be put into a one to one correspondence with the elements of a Heisenberg matrix. But since, according to Heisenberg, a different matrix corresponds to each quantity, it is necessary that we form for each quantity our combinations of proper functions in a different way.

Here, there arises an essential idea whose importance will appear even better to us in the next chapter. This essential idea is that it is necessary to make a certain symbol of operation, a certain operator, correspond to each physical quantity. We have already seen that, in order to form the propagation equation of the wave associated with a corpuscle by a somewhat automatic process, Schrödinger had been led to replacing the components of the momenta with operators proportional to derivatives with respect to the conjugate coordinates, the factor of proportionality containing the constant h. It is also natural to assume that to each coordinate there corresponds the operation "multiplication by this coordinate." Since all the mechanical quantities attached to a corpuscle can be expressed by means of the coordinates and the components of the momenta (conjugate moments of Lagrange), the two preceding rules permit us to find the operator corresponding to any mechanical quantity attached to the corpuscle. If the operator corresponding to the energy is formed, we find the Hamiltonian operator which served in forming the equation of propagation. Generalizing this correspondence, we set up the principle that all physical quantities are connected with an operator, and we make this principle one of the bases of the new mechanics.

And now we are in a position to understand how Schrödinger set up the matrices that he wanted to identify with those of quantum mechanics. Let there be a certain mechanical quantity attached to a corpuscle and the operator that corresponds to it, an operator that we know how to form. To each pair of proper functions of the system under consideration we can then associate a quantity formed in the following way: the operator in question is applied to one of the functions of the pair, the result is multiplied by the conjugate complex value of the other function and an integration over all space is performed. By repeating the same operation for all pairs of proper functions, an array of elements is obtained, some attached to a single stationary state, others to two stationary states, i.e., to a transition. These elements can be arranged in a table where the elements of the first kind are put on the diagonal (diagonal elements). Each mechanical quantity thus gives rise to a matrix and the question now is to find out if these matrices, as derived in wave mechanics, can be identified with those of quantum mechanics.

The answer to this question is in the affirmative. Schrödinger showed first of all that the matrices formed by the process just indicated do satisfy, as do those of Heisenberg, the rules of addition and multiplication of algebraic matrices. Furthermore, the somewhat strange way in which Planck's constant was introduced into quantum mechanics finds an immediate interpretation in the conceptions of Schrödinger. The product of two operators, in general, is not commutative as a matter of fact: the result obtained generally depends on the order of the operations. Nevertheless, in the greater number of cases, two operators corresponding to mechanical quantities are commutative: there is an exception, however, when these quantities are a coordinate and the conjugate component of the momentum, for the operator corresponding to this latter quantity is proportional to the derivative with respect to the conjugate coordinate, and the operation "derivation with respect to a variable" does not commute with the operation "multiplication by this variable" as can easily be seen. The non-commutation rules given by Heisenberg result immediately from this. Nothing more remains then in order to complete the identification than to show that the matrices formed by means of the wave functions obey the canonical equations of quantum mechanics. This is the way in which it is done: these canonical equations state precisely, as Schrödinger proved, that the wave functions, by means of which the matrices are constructed, do satisfy the equations of propagation of wave mechanics. In brief, the canonical equations of quantum mechanics are equivalent to the equation of propagation of wave mechanics.

Thus, the two forms of the new mechanics turn out to be reducible to each other, and it is not longer surprising that they lead to the same results in problems of quantization. The method of quantum mechanics, which works directly with matrices without passing through the intermediary of the wave functions, is more compact and often leads more quickly to the desired results. But the method of wave mechanics, better fitted to the intuition of physicists and more in agreement with their habits of thought, seems more natural at first, and easier to handle. Actually, more physicists make use of the wave method and make their calculations by using explicitly the wave functions.

4. The Correspondence Principle in the New Mechanics

THE NEW MECHANICS has permitted us to give the Correspondence Principle a much more precise form, and one less subject to the criticism to which it was susceptible in the framework of the old quantum theory. We have seen how Bohr had tried to use Fourier series developments of the electric moment corresponding in the classical picture to the initial or the final state of a quantum transition in order to predict the polarization and intensity of the radiation connected with this transition. In the case of large quantum numbers, the method develops satisfactorily, free from ambiguity; but in the case of average or small quantum numbers, the only important one in practice, difficulties and ambiguities are found. On the contrary, in the new mechanics, a thoroughly well-defined manner of applying the Correspondence Principle is at once obtained. Actually, a matrix corresponds to each component of the electric moment, one element of which matrix, and one only, is attached to each transition. By considering the matrix element which corresponds to a certain transition as the amplitude of the given component of the electric moment for this transition, a thoroughly precise and unequivocal prediction of the radiation connected with this transition can be obtained by making use of formulae patterned after classical formulae. To be sure, a somewhat hypothetical element remains in this method, which is the possibility of using formulae of a classical type in order to calculate the intensities, but this is one of the essential postulates of the method of correspondence. This hypothesis having been assumed, there is no longer anything unprecise or arbitrary in the application of the Correspondence Principle.

This precise form of the Correspondence Principle was stated by Heisenberg in his studies on matrix mechanics. It has been translated into the language of wave mechanics by Schrödinger. This eminent physicist has even proposed a kind of concrete picture to explain the role of the matrix elements in the calculation of radiation. In the atom, the electron might be considered as no longer being localized at each instant. It only has a certain probability of being found at such and such a point, which is proportional, according to the interference principle, to the square of the modulus of the wave function. This allows us to regard the electron as being somehow spread out in the atom, and its electric charge distributed on the average in a continuous fashion. According to Schrödinger, the Correspondence Principle might be applied by saying that things take place as if this mean distribution of electricity (which varies with the time) radiates according to classical laws. At first this interpretation seemed most satisfactory for it allows us to find a restatement of Bohr's law of frequencies, but when it is examined closely, one sees that it raises grave difficulties and that it must be rejected. In reality the process of emission by quantum transitions is too discontinuous in its essence for it to be capable of an accurate representation as an emission, realized according to classical laws, from any distribution of electricity, even fictitious. The only truly correct interpretation of the role of the matrix elements is to say in agreement with the ideas that we have stated apropros the Correspondence Principle: the matrix elements permit us to calculate the probability that any one state has of undergoing a certain quantum transition in a unit of time.

The Correspondence Principle of the new mechanics has permitted us to calculate the intensities and polarizations of spectral lines and, in particular, to recover the selection rules. It has also permitted us to study a great number of problems about the interactions between matter and radiation, among which I shall only cite the problem of the scattering of light and that of dispersion. We have been able to find exactly the formula of Kramers-Heisenberg, already obtained by the approximate considerations of correspondence.

The application of the method of correspondence to the study of the interaction between matter and radiation has given very satisfactory results and certainly holds a large element of truth. Nevertheless, it is impossible not to remark that by systematically employing the formulae of electromagnetism, suitably translated, it constantly overlooks the granular nature of light. In reality, the scattering of light by an atom ought to be treated as a collision problem between a photon and an atom, a collision studied, of course, by the methods of wave mechanics. In order to succeed in visualizing the question in this light, it is necessary to try to introduce photons into the electromagnetic wave and more generally to quantize the electromagnetic field. We shall have to speak again about the endeavors which have been made in this direction.

Chapter 10 | The Probability Interpretation of the New Mechanics

1. General Ideas and Fundamental Principles

We have already seen considerations of probability play a large role in the initial physical interpretations of wave mechanics. In these it was felt that a general theory was emerging which would attribute a probability character to all the predictions of the new mechanics. This theory, very new in aspect and upsetting many classical ideas, has by degrees thrust itself upon the attention of all physicists. It can be said that today it has been adopted by all, even by those who believe it provisional and who have not given up hope of returning someday to more classical conceptions. This is what we wish to discuss in this chapter.

We shall start with this idea, almost banal in appearance,

that in order to know the value of a physical quantity exactly it is necessary to measure it. And in order to measure it, there must always be a certain apparatus which somehow forces this quantity to make itself known with such and such a precise value. In classical physics, it was assumed *a priori* that, through appropriate precautions, it was always possible to make these measurements in a way so as not to disturb appreciably the state before the measurement. In these conditions, the measurement acts only to ascertain an existing state, it introduces no new element. On the macroscopic level, this postulate, implicitly admitted by classical physics, is correct. In this domain a capable experimenter can always study a phenomenon quantitatively without introducing a significant perturbation. This results from the fact that the perturbations produced by the operation of measurement can always be diminished sufficiently to make them negligible with respect to the quantities being measured. On the microscopic level, on the contrary, the existence of the quantum of action has as a consequence the fact that the perturbations arising in the operation of measurement can not be diminished indefinitely and, therefore, each measurement materially disturbs the phenomenon under study. These ideas will be stated precisely when we take up a little later the examples that have been given, principally by Bohr and Heisenberg, in support of the Uncertainty Relations. For the moment, it will be sufficient for us to note that it is in no way evident that an operation of measurement purely and simply makes the pre-existent state known to us: it might very well be that the operation of measurement results in the creation of a new state by extracting from the pre-existent state one of the possibilities contained in it. And now we must try to formulate precisely the role of measurement as it is conceived in this new light.

In order to accomplish this purpose, it will be pertinent to reflect a little on certain classical experiments in physical optics. Here again, it is by starting out from the duality of pho tons and light waves that we will have a better chance of being able to disentangle matters. We are therefore going to imagine a very usual experiment: the spectral analysis of a complex beam of light by means of a prism (or a grating). The effect of the apparatus used then is, as has been known since Newton, the separation of the various monochromatic components contained in the incident light. There was a great deal of discussion in the 19th century over the question of knowing if the monochromatic components isolated by the prism existed in the incident light or were created by the action of the prism. No very satisfactory answer had been given to this question, but in the end the more prudent attitude consisted in saying: the monochromatic components exist virtually, somehow in a potential state, in the incident light. We are going to see that this opinion is confirmed by analyses of a quantum nature of which we shall speak. We are going to try, as a matter of fact, to introduce the idea of photons into the interpretation of dispersion by a prism. From this point of view, we shall say that, through the action of the prism, the incident photons are separated into well-determined color groups: the prism extracts from the incident beam the red, yellow and blue photons. But we can imagine the experiment performed with beams that are so weak that the photons arrive one after the other at the prism. Each photon is associated with the incident light wave which, by hypothesis, is not monochromatic. A well-determined frequency therefore can not be attributed to the incident photon, nor consequently a well-determined energy by the Einsteinian relation; the incident photon has, somehow, several possible frequencies, viz., those which figure in the spectral decomposition of the associated light wave. But, upon leaving the prism, the incident photon can be met with in one of the monochromatic beams isolated by the action of the prism: hence, it now has a well-determined frequency. The prism therefore now appears to us to be an instrument permitting a measurement of the frequency (or the energy) of the photons: this apparatus has just the effect of extracting from the pre-existing state one of the possibilities contained in it. What we then must look for is an evaluation of the probability that the action of the prism will force the incident photon to have such and such a determined color. Wave theory immediately furnished us with a quantitative answer. The incident wave can be represented by a Fourier expansion where each monochromatic component possesses a definite amplitude. The effect of the prism is the separation of these monochromatic components while preserving their amplitudes, and the light energy which falls on the prism is divided at the outlet between the different emergent monochromatic beams, proportional to the squares of the amplitudes, to the intensities, of the various Fourier components. We must therefore say that the probabilities of a photon's possessing such a frequency after passing through the prism is proportional to the partial intensity corresponding to this frequency in the Fourier expansion for the incident light wave.

The above considerations, translated into the language of wave mechanics and generalized, permit us to understand the origin of the general probability theory which we are going to develop now.

The new mechanics, as we have seen in one of the last sections, makes a certain operator, that can in all cases be formed, correspond to each mechanical quantity. These operators in question all belong to the class of linear Hermitian operators. The mathematical theory of proper values, of which we have already had occasion to speak, permits us to make proper values and proper functions correspond to these operators. Because the operators are Hermitian, the proper values are real constants forming a continuous, discontinuous or "mixed" sequence which makes up the "spectrum" of the operator. The proper functions form, at least in the general case, a complete set of orthogonal functions, i.e., any continuous function whatsoever can be expanded in a series of these proper functions. These properties of proper values and proper functions, we have already encountered for the proper values and proper functions of the Hamiltonian operator in Schrödinger's method of quantization. In this method, it is assumed, as we have seen, that the only possible values of the energy of a quantized system are the proper values of the Hamiltonian operator which corresponds to its energy. Generalizing this idea, the general probability theory of wave mechanics assumes the following first fundamental postulate which can be called the "principle of quantization":

"The exact measurement of a mechanical quantity can furnish as the value of this quantity only one of the proper values of the corresponding operator."

In each case, this postulate fixes the possible values of a mechanical quantity. It obviously must be complemented by a second postulate telling us, for a corpuscle whose initial state prior to the measurement is known, what are the respective probabilities of the various possible values, i.e., the probabilities of the various possible results of the measurement. Now, the initial state of a corpuscle prior to the measurement supposedly known, is represented in wave mechanics by a certain ψ-wave. It is this ψ-wave which is going to strike the measuring device. The analogy with spectrum analysis by a prism then tells us the second postulate to be adopted. The ψ-wave can actually be resolved into a series of proper functions corresponding to the physical quantity to be measured. We are quite naturally led to think that the squares of the amplitudes of the components in this spectral resolution will measure the relative probabilities of the various possible values. So we can state a second fundamental postulate that might be called "the generalized principle of spectral resolution":

"The respective probabilities of the different possible values of a mechanical quantity attached to a corpuscle for which the ψ-wave is known are proportional to the squares[1] of the amplitudes of the corresponding components of the spectral resolu-

[1]More correctly "to the squares of the moduli."

tion of the ψ-wave into proper functions of the quantity considered."

It is most evident that this second principle contains as a particular case the principle of spectral resolution due to Born of which we have already spoken and which applies to the quantity "energy." It is a great deal less evident that this principle also contains as a special case what we have already called the interference principle. However, an argument, which I can not reproduce here, shows that by applying the generalized principle of spectral resolution to the quantities "coordinates of a corpuscle" we obtain the interference principle. Thus, the two principles which we had introduced in the chapter before last in order to inaugurate the physical interpretation of wave mechanics prove to be special cases of the second fundamental postulate of the general theory. The two fundamental postulates that we have stated in the present section are therefore sufficient to serve as the basis for a complete and coherent probability interpretation of the new mechanics. Obviously, there are small secondary questions concerning which we can not speak here: in order to obtain the absolute value of the probabilities it is necessary to "normalize" the proper functions and the ψ-function; in order to take account of degenerate cases where there are multiple proper values, it is necessary to expand the statement of the second postulate, etc. But these are just details and the body of the theory is established in a logically satisfactory fashion.

And now, we wish to anticipate an objection that more than one reader must have expressed while reading this. This probability interpretation of the new mechanics, it no doubt will be said, is perhaps very nice and very coherent, but is it not a little arbitrary? Why go out and seek conceptions so complicated and so contrary to the usages of classical mechanics? To this we answer that the probability interpretation of which we have just sketched the outlines today seems to be the only possible one. By this we mean that it alone permits us at this time to account for the whole of the quantum phenomena in the framework of wave mechanics, as it is imposed by experiment. No attempt made in any other direction has been able to succeed; the author of this book knows this better than most for he has made attempts of this kind which he has finally had to give up by reason of the insurmountable difficulties encountered.

To sum up, the fundamental postulates stated above are justified by the possibility of founding on them a coherent theory, compatible with all the experimental facts, and by the impossibility of finding another system that possesses these same qualities. In reality, all physical theories are always justified by reasons of this kind, for at the base of any physical theory lie arbitrary postulates, and it is the success of these postulates that makes their use legitimate.

We shall state precisely in the following sections the deep differences which separate the probability interpretation of the new mechanics from classical theories and we shall limit ourselves here to pointing out that the principles studied in this section have taken on a more abstract and even more general form in the works of scientists such as Dirac and Jordan under the name of the theory of transformations. The deeply mathematical character of these developments will not allow us to treat them here.

2. *The Uncertainty Relations*

THE PHYSICAL interpretation of the new mechanics leads to some most interesting and important consequences to which Heisenberg was the first to draw attention. They are mathematically expressed by the inequalities that are familiar today under the name of the Uncertainty Relations. Heisenberg had proved these inequalities by relying on the non-commutation relations of his new quantum mechanics. To explain their meaning we shall rely here on the picture that wave mechanics furnishes us: we shall show that they necessarily result from the physical interpretation previously assumed for this mechanics if it is supposed that the state of a corpuscle is always representable by a ψ-wave.

First, let us consider a plane monochromatic wave associated with a free corpuscle. We know that to this wave there corresponds a completely determined state of motion, hence a well-defined vector, "momentum." This is what we express by saying that the state under consideration constitutes a "pure case" (Reine Fall) for the momentum and consequently for the energy. But a plane monochromatic wave has everywhere a constant amplitude and the interference principle then obliges us to say that the position of the corpuscle is completely indeterminate, its probability of being found at no matter what point in space is the same. So we assert that a complete determination of the state of motion of a corpuscle implies, according to the principles of the new mechanics, total indeterminancy of its position in space. But, naturally, the case when the ψ-wave associated with a free corpuscle is plane and monochromatic is a very special case: in general this ψ-wave will form a wave packet formed by the superimposition of a certain number of plane monochromatic waves. The dimensions of the wave packet can then be placed within limits and the position of the corpuscle will be better determined since it necessarily will be found in the region occupied by the wave packet, the only region where the amplitude will be different from zero. Here there enters a property of the mathematical representation of a wave packet by an expansion in Fourier integrals. This property is that, the smaller the dimensions of a wave packet, the more extended is the spectral interval occupied by the components of its Fourier resolution. In more suggestive terms, we might say: the less extended a wave packet is, the less it is monochromatic. It then becomes obvious, by invoking the two principles of interference and spectral resolution, that the state of motion of a corpuscle is that much more uncertain when its position is better defined. What we gain on one hand, we lose on the other. Finally, we can pass to the limiting case which is the counterpart to that of the plane monochromatic wave. To do this, we imagine a wave packet ψ of infinitely small dimensions. The position of the associated corpuscle is then exactly known: we are dealing with a pure case for the position. But in this limiting case, the representation of the wave packet can be made only by a Fourier integral extending over all possible plane monochromatic waves, and our fundamental principles oblige us to say that the state of motion is completely indeterminate. Thus a precise knowledge of position implies a complete ignorance of the state of motion. Our general conclusion is then that the fundamental principles put at the base of the physical interpretation of wave mechanics, together with the mechanism of the representation of a wave packet by a superimposition of monochromatic waves, implies the impossibility of knowing with precision and at the same time the position and state of motion of a corpuscle.

We have just given, in a rather qualitative way so as to make it a little easier to understand, the reasoning which leads to the Uncertainty Relations of Heisenberg. Developed more rigorously, his reasoning leads to the following result: the product of the uncertainty of a coordinate by the uncertainty of the corresponding component of momentum is always at least of the order of magnitude of Planck's constant h. Thus the stated inequalities are obtained. They show that a coordinate of a corpuscle and the corresponding component of momentum can not be known at the same time with precision and that, if the uncertainty about one of the two conjugate quantities is very small, the uncertainty about the other is very large.

The Uncertainty Relations, let us repeat, are a necessary consequence on one hand of the possibility of making the state of a corpuscle correspond to a certain associated wave, and on the other hand, of the general principles of the probability interpretation. But it is still necessary, after these arguments have been made, to show that no measurement can ever lead to a knowledge of the position and motion of a corpuscle with greater precision than that which the Uncertainty Relations permit; without that, it would prove impossible to be able always to represent the state of a corpuscle by a certain associated wave. Keen and profound analyses of the procedure of measurement have permitted Heisenberg and Bohr to show that no measurement can effectively lead to results in disagreement with the Uncertainty Relations. And this, as we are going to see, is bound up with the existence of two essential discontinuities which most probably are related to each other: the quantum of action on one hand and the discontinuous structure of matter and radiation on the other.

To understand why experiment can not furnish us with greater precision than the Uncertainty Relations will allow, let us imagine that we are trying to localize a corpuscle with precision. The most delicate way we have at our disposal for exploring space on a very small scale, is to use radiation of short wave length. This method, much more precise than any mechanical method, will allow us to distinguish two points in space whose distance is at least of the order of this wave length. To determine the exact position of a corpuscle, we must use radiation whose wave length will have to be that much shorter as we desire more precision. But here the existence of the quantum of action enters in the form of radiation quanta. The more we diminish the wave length of the exploratory radiation, the more we increase its frequency and consequently the energy of its photons, as well as the momentum that these photons can transfer to the corpuscle being studied. The measuring apparatus, directed to an exact determination of the position, will leave us ignorant of the change of momentum experienced by the corpuscle during the measurement, so that the final state of motion of the corpuscle, after the measurement has been made, will be that much more uncertain when its position has been made more precise. And by making the preceding discussion quantitative, the Uncertainty Relations are found once again. Inversely, measurements relating to the state of motion can be imagined: for example, a determination of the velocity of an electron can be attempted by examining the light it scatters with Doppler effect. We again come to the conclusion that the more precise the instrument fixes the state of motion of a particle, the more uncertain is the position of the latter once the measurement has been made and the Uncertainty Relations again are a mathematical translation of this fact. We can not develop here in detail the numerous examples which have been given by Bohr, Heisenberg and others, for that would demand figures and formulae. These examples are convincing and it seems that today most physicists admit the impossibility of finding a measuring instrument allowing an infringement of the restrictions contained in the inequalities of Heisenberg.

Before examining certain philosophic aspects of the results given in these last two sections, we would like to show first of all why the Uncertainty Relations, and more generally, the general principles of the probability interpretation as given above, are not in contradiction with the verified predictions of the old mechanics, but on the contrary admit them as being valid as first approximations.

3. The Accord with the Old Mechanics

FROM THE BEGINNING of the development of the quantum theory, it had been evident that, if classical mechanics is not rigorously exact, the responsibility for this rests on the existence of the quantum of action. In other words, if Planck's constant had a zero value, classical mechanics would be exact. In all branches of the old quantum theory, from Planck's theory of black body radiation to the extreme developments of the ideas of Bohr and Sommerfeld, everywhere we find that by making the value of h tend to zero, the quantum formulae come into coincidence with classical formulae.

This fundamental idea is found again in the new mechanics. If we take the point of view of quantum mechanics, all differences between the old and the new mechanics arise from the non-commutation of the matrix corresponding to each coordinate with the matrix corresponding to the conjugate Lagrangian moment and the lack of commutation being proportional to h would disappear if h were zero. If we prefer to take the point of view of wave mechanics, it will be noted that, the wave length of the ψ-waves being proportional to h would be zero if h vanishes and that then geometrical optics would always be valid, for it is easy to see that geometrical optics is always applicable when the wave length is infinitely small. Therefore with h tending to zero, the equation of propagation of the ψ-wave could always be replaced by the equation of geometrical optics, i.e., as we know, by the Jacobi equation. So an asymptotic accord between the new mechanics and the old would be realized.

Hence it is very easy to understand why classical mechanics is entirely valid in practice for large scale phenomena, macroscopic phenomena. These phenomena, in effect, bring into play such large values of physical quantities that the quantum of action can be considered to be absolutely negligible in them and the effects of its existence to be completely masked by the inevitable lack of precision in physical measurement. It is easy to make this clear by numerical examples and to show, for example, that, in order to confirm Heisenberg's inequalities for a ball of a tenth of a milligram (an exceptionally favorable case because of the great lightness of the ball), it would be necessary, even if the velocity were known to within one millimeter per second, to be able to measure the position of its center of gravity to within less than 10^{-20} centimeter! But in order to understand even better how the agreement between the old and the new mechanics is made, we are going to examine a specific case a little more closely.

Let us suppose that we are studying the motion of a corpuscle on a large scale, for example, the motion of an electron in a magnetic field. We know that we can exactly describe this motion by means of the conceptions of classical mechanics. How is this in accord with the Uncertainty Relations? The starting point for an explanation is the following remark: in the conditions of this macroscopic experiment, the smallest distance that we would be able to measure directly is a great deal larger than the wave length of the waves associated with the corpuscle being studied. Consequently, there can exist a wave packet whose dimensions are smaller than what we can measure directly and which, nevertheless, will be formed by waves of almost identical wave lengths. Therefore a well-run and precise experiment can permit us, without contradicting Heisenberg's relations, to represent the state of the corpuscle after the measurement by a "wave group." Since this group is practically point-like and practically monochromatic for us, we will be able, within the limits of precision of macroscopic measurement, to attribute a well-determined position and velocity to the corpuscle. Furthermore, a fundamental result obtained at the very beginning of wave mechanics tells us that a group of ψ-waves is displaced at each instant with a velocity that classical mechanics would assign to the associated corpuscle. Therefore our quasi-point-like group of waves would move exactly like a classical corpuscle and since, according to the interference principle the actual corpuscle must always be within the wave packet, everything happens as if the actual corpuscle obeyed the laws of classical mechanics. As we can realize from this example, it is just the lack of precision in our macroscopic measurements which masks the quantum uncertainties.

There does not seem to be then any serious difficulty in the agreement between the new mechanics and the old. The structure of quantum physics would seem to be built around classical physics and to have enveloped it within a wider framework. So, in all the long history of science, progress is made by successive approximations.

4. *Indeterminism in the New Mechanics*

THE EQUATIONS of classical mechanics completely determine the motion of a system when the position and state of motion of the parts of the system are known at the initial instant. Thus the classical motion of a corpuscle is completely predictable when its position and velocity are known at a certain initial instant. This possibility of predicting in an inexorable way the future of a mechanical system when a certain amount of data about its present state is known constitutes the determinism of classical mechanics. The striking successes won by this mechanics, notably in mathematical astronomy, had led all physicists to try to set up a theoretical physics where determinism would always be verified. The macroscopic phenomena that they then studied were bent to this demand and all of classical theoretical physics rested on equations in derivatives or partial derivatives, permitting a rigorous calculation of the evolution of any physical system whatsoever, starting out with certain data on its initial state. Even in the branches of physics where the calculus of probabilities had been introduced, it was always supposed that the elementary phenomena obeyed a rigorous determinism and that only the very large number and randomness of the elementary phenomena contained in the over-all phenomenon being studied permitted an appeal to statistical methods and the notion of probability. More or less consciously, the inner determinism of natural phenomena, implying their complete predictability at least in principle, had become a kind of scientific dogma. We are going to see that the development of the new quantum theories has profoundly modified this situation.

We can realize the difference which exists on this score between the old and the new mechanics by noting that the elements whose simultaneous knowledge at the initial instant was necessary in classical mechanics in order to be able to predict rigorously the evolution of a system, are precisely those whose simultaneous determination is impossible according to the Uncertainty Relations. As we have recalled before, in order to solve rigorously the classical equations of motion for a system, it is necessary to know its configuration and the state of motion of its parts at a certain instant. Since any system is reducible in the last analysis, in the eyes of modern physics, to an assemblage of corpuscles, it would therefore be necessary to know the coordinates and velocities (or momenta) of the various corpuscles of the system at the same instant. Now the essential content of the Uncertainty Relations is just in declaring such precise and simultaneous knowledge impossible. Assuredly, the order of magnitude of the constant h, an extremely small constant with respect to our usual units, makes the quantum uncertainties negligible for physical phenomena on an ordinary scale and determinism apparently rigorous. But, in the microscopic study of physical phenomena, the importance of the uncertainties will be considerable, and sufficient to forbid completely a description of the course of events consistent with the demands of determinism.

The disappearance, or at least relaxation, of determinism in quantum physics has been balanced by the appearance of laws of probability. But the appeal to probabilities has a very different significance here than it had in statistical mechanics for example. In classical theories where probabilities enter, the elementary processes were considered to be controlled by rigorous laws, and probabilities were introduced to describe large-scale phenomena relating to an immense number of elementary phenomena. In quantum physics, on the contrary, probabilities are directly introduced to describe the course of elementary processes. In order to understand better how the question presents itself, we must show exactly how the new mechanics represents the evolution of elementary phenomena by means of waves.

We shall base our argument on a single corpuscle. Our considerations could easily be extended to a system of corpuscles by the method which will be applied in Chapter 12.

Knowing the result of a certain number of observations or

experiments, the aim of theoretical physics is to predict the result of other observations or forthcoming experiments. In classical physics, it is assumed that it is possible to measure at the same instant the coordinates and the velocity of a corpuscle and the equations of classical dynamics then allow us, in principle, to predict rigorously the result of an observation or measurement made on this corpuscle at a later time. In the new mechanics, we are, on the contrary, led to assume the impossibility of measuring, simultaneously and precisely, the coordinates and the momentum of the corpuscle. A measurement, even if made with the greatest precision experimentally possible, can not furnish for these quantities information affected with smaller uncertainties than those imposed by the inequalities of Heisenberg. The state of a corpuscle, such as it is known after a measurement, will be represented by an associated wave which will never be both point-like and monochromatic: it will always have a certain extension either in space or in a range of frequencies, and generally even in both at once. The equation of propagation then allows us, starting from this initially known form of the ψ-wave, to calculate exactly the evolution of the wave during the entire period when no new observation or measurement is made, and consequently to state, at each instant, the probability of finding such and such a value for such and such a quantity attached to the corpuscle, if a measurement permitting a determination of it were made at this moment. When a new measurement will have effectively been made, it will furnish new knowledge on the state of the corpuscle and will completely overthrow the situation as to the probabilities, just as the situation about the probability of an event is upset when information about this event is obtained. It will therefore be necessary, after this new measurement, to construct a new ψ-wave which will represent the new state of our knowledge relative to the corpuscle. Taking up again the idea developed at the beginning of this chapter, we will say that each experiment provokes, as a consequence of the existence of the quantum of action, an uncontrollable disturbance in the state of the corpuscle which will not permit an exact causal relation to be established between the prior and subsequent state. This disturbance is connected with the existence of the quantum of action because, as we have seen, particularly in the last section, it is this which stands in the way of the unlimited reduction of the causes of uncertainty in the operation of measurement. The evolution of the ψ-wave between two measurements is completely determined by its initial form and by the equation of propagation: it consequently obeys a rigorous determinism, but it by no means follows from this that there is a rigorous determinism of observable and measurable phenomena, each new observation and measurement acting to add new elements and to disrupt the regular evolution of the ψ-wave.

Heisenberg has given an example of the application of the preceding considerations. He pictures two successive measurements of the position of a corpuscle. The first measurement permits the corpuscle to be localized in a small region of space. The associated wave after this first measurement will then be a wave packet limited to this region of space (without this we would not be in agreement with the interference principle). This wave packet, which perforce will be far from being monochromatic, will expand by propagating itself as the equation of propagation shows. The second measurement of the position, made at a certain later instant, will permit the corpuscle to be located in a new small region of space which necessarily will be within the region occupied at this instant by the wave packet and which, in general, will be a great deal smaller. In other words, as a consequence of the wave propagation, the field of possible positions of the corpuscle increases very rapidly, and the effect of the second measurement is to reduce it brusquely. After the second measurement, it will be necessary to construct a new wave packet ψ with dimensions a great deal more reduced than those of the first one in its final state. This new form of the ψ-wave will naturally be the point of departure for a new evolution of probabilities.

Now we can understand how the conceptions of the new quantum physics have shattered the old pretensions of determinism. Obviously, there still are cases where the result of a measurement, the value of a quantity, can be predicted with certainty; this is when the state prior to a measurement constitutes a "pure case" for this quantity, or, in other words, when the expansion of the ψ-function in proper functions corresponding to this quantity, is reduced to a single term. Such is the case for a measurement of energy or momentum made on a corpuscle which is associated with a plane monochromatic wave. But these are exceptional cases, of which it might even be said that, strictly speaking, the probability is zero.

There has been a great deal of discussion in the last years about this question of indeterminism in the new mechanics. A certain number of physicists still manifest the greatest repugnance to consider as final the abandonment of a rigorous determinism, as present day quantum physics must do. They have gone to the length of saying that a non-deterministic science is inconceivable. This opinion seems exaggerated to us, since quantum physics does exist and it is indeterministic. But it seems to us perfectly permissible to think that, some day or other, physics will return to the paths of determinism and that then the present stage of this science will seem to us to have been a momentary detour during which the insufficiency of our conceptions had forced us to abandon provisionally our following exactly the determinism of phenomena on the atomic scale. It is possible that our present inability to follow the thread of causality in the microscopic world is due to our using concepts such as those of corpuscles, space, time, etc.: these concepts that we have constructed by starting with the data of our current macroscopic experience, these we have carried over into the microscopic description and nothing assures us, but rather to the contrary, that they are adapted to representing reality in this field. Nevertheless, although a great many fundamental reforms still seem necessary to us in order to have a clear understanding of quantum physics, personally it does not seem probable that we will succeed in completely re-establishing the determinism of days gone by. The blows that the development of the new mechanics have delivered to it seem to us too deep-seated to be easily effaced. The wisest course, no doubt, is to hold to this statement: at the present time, the physics of phenomena where quanta play a part, is no longer deterministic.[1]

5. *Complementarity, Idealization, Space and Time*

Bohr, whose role has been so essential in the whole development of contemporary physics, has contributed a great deal in his studies, which are always profound and often subtle, toward clarifying the meaning of the orientation, which is so original, of the new mechanics. It is he, in particular, who introduced the notion of complementarity which is so curious from a philosophical point of view.

Bohr starts with this idea that the description of an entity like an electron must be made as much with the help of the corpuscular picture as with the wave picture, and wonders how two so different pictures, so contradictory one might say, can thus be used concurrently. He shows that this can be done because the Uncertainty Relations, consequences of the existence of the quantum of action, do not permit the two images to enter into direct conflict. The more precise it is desired to make one picture through observations, the hazier the other one becomes. When the electron has a wave length sharply enough defined so that it can interfere with itself, it is not localized and does not respond any longer to the corpuscular picture; on the contrary, when the electron is definitely

[1] J. von Newmann has proven that the probability laws of the new mechanics are incompatible with the existence of a hidden determinism, which makes it most improbable that determinism in atomic physics will be re-established in the future.

localized, its interference properties disappear and it does not respond any longer to the wave picture. The wave and corpuscular properties never enter into conflict because they never exist at the same time. We are continually expecting a battle between the wave and the corpuscle: it never occurs because there is never but one adversary present. The entity "electron," as well as the other elementary entities of physics, thus has two irreconcilable aspects, which however must be invoked in turn in order to explain all of its properties. They are like the two faces of an object that never can be seen at the same time but which must be visualized in turn, however, in order to describe the object completely. These two aspects, Bohr calls them "complementary" aspects, meaning by that that these aspects on one hand contradict each other, and on the other complement each other. And, in its essence, this notion of complementarity seems to have taken on the importance of a true philosophical doctrine.

Actually, it is in no way evident that we could describe a physical entity by means of a single picture or of a single concept of our intellect. We construct our pictures and our concepts, drawing our inspiration from our daily experience; we extract from this experience certain aspects and, starting from there, we forge through simplification and abstraction certain simple pictures, certain apparently clear concepts which we finally try to use in interpreting phenomena: such are the concepts of a well-localized corpuscle, of a strictly monochromatic wave. But it may be that these "idealizations," as Bohr says, the overly simplified and highly rigid products of our minds, can never be applied exactly to reality. In order to describe the complexity of reality, it may accordingly be necessary to use successively two (or several) of these idealizations for a single entity. Now one, now the other will be more adequate: sometimes (in the pure case of the preceding section) one of the two will be exactly adapted to the description of the entity being considered, but this case will be most exceptional and, in general, we will not be able to avoid making an appeal to two ideal images.

If we correctly penetrate the complex thought of this illustrious physicist, such are some of the truly original considerations that quantum physics has inspired in Bohr. One might try to extend the field of application of these philosophical ideas beyond the domain of physics. We might, for example, investigate, as Bohr himself has done, if the notion of complementarity can not find important applications in biology and aid in understanding the physico-chemical and characteristically vital double aspect of the phenomena of life. We also could, in another range of ideas, examine whether all "idealizations" are not that much less applicable to reality when they become more complete and, although we have little inclination to be paradoxical, we could hold, contrary to Descartes, that nothing is more misleading than a clear and distinct idea. But it is wise to stop at this dangerous point and return to physics.

What is more certain is that our customary notions of space and time, even when profoundly modified by the theory of relativity, are not strictly appropriate for a description of atomic phenomena. We have already noted, principally in the introduction, that the existence of the quantum of action implies a quite unforeseen relation between geometry and dynamics: the possibility of localizing physical entities in the geometrical framework of space and time proves to be not independent of their dynamic state. Without doubt general relativity has already taught us to consider the local properties of space-time as being dependent on the distribution of matter in the universe. But the modification that the presence of quanta requires is even much more profound and no longer permits us to represent the motion of a physical unit by a line in space-time (world-line); it no longer permits us to define the state of motion starting from the curve which represents successive localizations in space in the course of time and demands that we consider the dynamic state, no longer as being derived from spatio-temporal localizations, but as an independent and complementary aspect of physical reality.

In truth, the notions of space and time drawn from our daily experience are valid only for large scale phenomena. It would be necessary to substitute for them, as fundamental notions valid in microphysics, other conceptions which should lead to our finding asymptotically, when we go from the elementary phenomena to observable phenomena on the ordinary scale, the customary notions of space and time. Need we say that this is a difficult task? We might even wonder if it is possible that we could ever succeed in eliminating from this matter something which constitutes the very framework of our everyday life. But the history of science shows the extreme productivity of the human mind and we must not give up hope. However, as long as we have not yet succeeded in expanding our concepts in the indicated direction, we shall have to strive to make microscopic phenomena enter, more or less awkwardly, into the framework of space and time and we shall have the painful feeling of trying to enclose a jewel in a setting which was not made for it.

6. *Will Quantum Physics Remain Indeterministic?*

In an article which appeared in the *Revue de Metaphysique et de Morale* under the title "Personal Recollections of the Beginnings of Wave Mechanics," and which subsequently was reprinted in my book *Physique et Microphysique,* I recounted the mental states I had passed through between 1923 and 1928 in so far as the interpretation of wave mechanics was concerned, and I explained that, after having tried to develop a concrete and deterministic interpretation consistent, in its larger aspects, with the traditional conceptions of physics, finally, in the face of difficulties which I met and objections which were made to me, I had to align myself with the probabilistic and indeterministic point of view of Bohr and Heisenberg. For almost 25 years now, I have remained faithful to this point of view, which, in addition, has been adopted by almost all theoretical physicists, and I have held to it in my teaching, my lectures and my books. In the summer of 1951, I received knowledge, through a friendly personal communication from the author, of a paper by a young American physicist, Mr. David Bohm, a paper which subsequently appeared in the January 15, 1952 issue of the *Physical Review.* In this paper, Mr. Bohm takes up in their entirety my conceptions of 1927, at least in one of the forms which I had given to them, meanwhile completing them in an interesting fashion in respect to several points. Then J. P. Vigier called my attention to the resemblance which exists between a demonstration given by Einstein concerning the motions of particles in general relativity, and a demonstration that I had given quite independently in 1927 in an hypothesis I had called the "theory of the double solution." All of these circumstances have redirected my attention to these questions, and, without being willing to assert that it is possible to re-establish a deterministic conception of wave mechanics in the meaning of my original ideas I believe, that the question is worth re-examining. We must, however, beware of all preconceived philosophic ideas and concern ourselves only with learning whether a coherent interpretation of all the well-established facts might thus be obtained.

Around 1920, at the time when I returned to scientific research after a long period of war service, the situation was as follows. On one hand, the existence of photons, soon to receive new confirmation in the discovery of the Compton and Raman effects, seemed certain; but the need to invoke the wave theory, in order to introduce the frequency ν, which figures in the definition of the photon, and also in order to account for the whole body of interference and diffraction phenomena—the laws for which had been established with great precision—showed the necessity of a synthetic viewpoint, expressing itself in a wave-corpuscle duality for light. On the other hand, the existence of quantized motions for corpuscles on a microscopic scale, suggested the idea of introducing the wave-corpuscle duality for electrons and other material elements, too. It seemed obvious to me, therefore, that it was necessary to bring about a general synthesis applicable to matter as well as to light, connecting, in formulae where the Planck constant h would necessarily figure, the wave and corpuscular aspects which are so indissolubly linked together.

It is this synthesis whose foundation I set forth in some notes which appeared in the *Comptes Rendus* of the Academy of Sciences at the beginning of the autumn of 1923 and, in a more complete fashion, in my doctor's thesis submitted in November of 1924. Taking inspiration from relativity considerations and also from ideas related to those that Hamilton had developed in the preceding century, I succeeded in associating with the motion of any corpuscle the propagation of a wave whose frequency and wave length were connected with the energy and momentum of the corpuscle by formulae where the constant h figures;[1] and I showed that one could, in this way, understand the reason for the existence of the quantized motions of the electrons in atoms. Without entering here into any technical detail, I shall emphasize the following point. With the rectilinear and uniform motion of a corpuscle in the absence of a field I associated the propagation of a plane monochromatic wave in the direction of motion having a constant amplitude and a phase linear in x, y, z, t. Since I established a relationship between the corpuscle's energy and momentum on one hand, and the wave's frequency and wave length on the other, I connected the state of motion of the corpuscle with the phase of the wave. But how should we make the fact that the corpuscle is localized, and has a position in space, correspond with this wave? A difficult question to resolve, for the plane monochromatic wave, having the same amplitude at any point

[1] See Chap. 8, Sect. 2 for a more complete statement of the author's original ideas on wave mechanics. Tr.

in space, in no way permits a definition at each instant of a privileged point which would be the position of the corpuscle at this instant. This difficulty, joined with several other relativistic considerations which I shall pass over, made me think that, if the phase of the plane monochromatic wave has a certain physical meaning, it is not the same for the constant amplitude of this wave: the uniform distribution of this amplitude in space would simply signify that *a priori* the corpuscle might be found at any point of space whatsoever with an equal probability. The amplitude would therefore have only a probability meaning, and the true position of the corpuscle (for I did not then doubt that this position must exist at each instant) would not be represented by it. I had therefore given to the wave which I introduced the name of the "phase-wave" so as to show clearly that in my eyes it was essentially the phase of this wave which possessed a physical meaning.

In the period which lasted from the submission of my thesis in November of 1924 to the session of the 5th Solvay Congress on Physics in October of 1927 I naturally followed with intense interest all the successive stages in the development of wave mechanics. But I was constantly worried by the question of the physical interpretation of the formalism of the new theory, and of the actual meaning of the wave-corpuscle dualism. To my knowledge, three possible interpretations of this dualism have been advanced. One interpretation, which always seemed to be preferred by Schrödinger, consists in denying the reality of the dualism by disputing the existence of corpuscles. Only the wave would have a physical significance analogous to the wave of classical theory. In certain cases, the propagation of the waves will give rise to corpuscular appearances, but they would be only appearances. At the beginning, to make his ideas precise, Schrödinger wanted to compare the corpuscle to a small wave train, but this interpretation can not be upheld, because a wave train always has a tendency to expand rapidly and continually in space, and consequently could not represent a corpuscle endowed with a protracted stability. Although Schrödinger would still seem to hold to interpretations of this type, I do not believe, on my part, that they are acceptable, and I think that it is necessary to assume the wave-corpuscle duality as a physical fact. Now, it is precisely the two other interpretations to which I alluded that assume this duality to be real, but they see it from two quite different points of view.

The first, to which I remained committed until 1928, consists in giving the wave-corpuscle duality a concrete significance consistent with the traditional ideas of physics, interpreting it for this purpose by considering the corpuscle as a kind of singularity in the midst of an extended wave phenomenon whose center it would be. The difficulty here exists in understanding why wave mechanics makes use of *continuous* waves without singularities of the type of continuous waves used in the classical theory of light. I shall say shortly in what form I tried to develop this viewpoint.

The second interpretation of the wave-corpuscle dualism consists in considering only the ideas of corpuscle and continuous wave, regarding them as "complementary sides of reality" in the sense that Bohr gives to this expression.[1]

In 1924, just before I submitted my thesis, I was completely imbued with the concepts of classical physics and it was within the framework of these concepts, within the framework of a Cartesian representation of phenomena by configurations and motions, that I sought to interpret the new ideas which I had introduced. It seemed certain to me that the corpuscle must have at each instant a position in space and a velocity; consequently, that it described a trajectory in the course of time. But I was convinced too that it was connected with a periodic and undulatory phenomenon permitting a frequency and associated wave length to be defined. I therefore quite nat-

[1] This third point of view is outlined in detail in Chap 10, Sects. 1 and 4. Tr.

urally imagined the corpuscle to be a kind of singularity in the midst of an extended wave phenomenon, the two forming a single physical reality. Since the motion of the singularity was connected with the evolution of the wave phenomenon at whose center it was, it would depend on all the circumstances that the wave phenomenon would encounter in its propagation through space. For this reason, the motion of the corpuscle would not follow the laws of classical mechanics, which is a purely point mechanics where the corpuscle undergoes only the action of the forces which act on it along its trajectory without suffering any repercussion from the existence of obstacles that can lie far outside its trajectory. In my conception, on the contrary, the motion of the singularity would undergo the influence of all the obstacles which would influence the propagation of the wave phenomenon with which it is linked, and thus the existence of interference and diffraction would be explained.

But then the difficulty lies in understanding why wave mechanics had developed by dealing solely with continuous solutions, without singularities, for the propagation equations, solutions which are usually designated by the Greek letter ψ. As I have already said, when I had associated the propagation of a wave, the plane, monochromatic wave ψ, with the rectilinear and uniform motion of a corpuscle, I ran into this difficulty: the phase of the wave, which permits a definition of the frequency and wave length of the wave associated with the corpuscle, would seem to have a direct physical meaning, while the constant amplitude of the wave, in my eyes, could only be a statistical representation of possible positions of the corpuscle. Here we had a mixture of the individual and statistical which intrigued me and seemed in urgent need of clarification.

If one refers to the notes which I published from 1924 to 1927 on this subject, one would see my thinking orienting itself by degrees to what I called then the "theory of the double solution." I gave an overall exposition of it in an article in the June 1927 issue of the *Journal de Physique* (v. 8, 1927,

p. 225) which remains the only complete document on this question. In this paper, I boldly postulated that any continuous solution ψ of the equations of wave mechanics was somehow doubled by a solution u with a singularity, this singularity in general is not fixed (the corpuscle), and with the same phase as the solution ψ. The two solutions ψ and u would thus both have the form of a wave, the phase being the same function of x, y, z, t, but the amplitude being quite different since that of u has a singularity while that of ψ would be continuous. Starting from the propagation equation, supposed the same for u and ψ, I then proved a fundamental theorem: the moving singularity of u should, in the course of time, describe such a trajectory that, at each point, its velocity would be proportional to the gradient of the phase. In this way, one could say, the reaction of the propagation of the wave phenomenon, on the singularity which forms its center, is introduced. I also showed that this reaction could be expressed by considering the corpuscle-singularity as being subject to a quantum potential which was precisely the mathematical expression for the reaction of the wave on itself. I thus took over an idea of the protagonists of the old corpuscular theory of light who said that in the diffraction of light at the edge of a screen, the light corpuscle undergoes an action from the edge of the screen, and is consequently deviated from its rectilinear path.

Since the u wave with its moving singularity in this way constitutes the corpuscle, and the undulatory phenomenon which surrounds it, what was the meaning of the ψ-wave? For me, it had no real physical significance, the physical reality being described by the u wave. But since the ψ-wave was supposed to have the same phase as the u wave and since the corpuscle-singularity always moved along the gradient of this phase, the possible trajectories of the corpuscle coincided with curves orthogonal to the surfaces of equal phase of ψ and I then showed simply that this would lead to considering the probability of finding the corpuscle at a point as being equal to the square of the amplitude, to the intensity of the ψ-wave.

Such was the subtle and curious interpretation of wave mechanics that I tried to develop in 1927. It did not take long to realize that its justification would run into very serious mathematical difficulties. For it was necessary to show that, in a well-defined problem in wave mechanics, with its boundary conditions and where the solution of the ψ-type was known, there also existed solutions of the u type with a moving singularity. It was also necessary to recast the theory of interference phenomena using *solely* the u wave with a singularity, the only physical reality, without bringing in the continuous wave, considered fictitious. It was necessary to interpret by means of the u waves the wave mechanics of systems of corpuscles developed by Schrödinger in the framework of the space of configuration, etc. . . . But I did not feel myself capable of resolving these difficult mathematical problems involving the arduous study of solutions with singularities.

At the present time, the new examination I have just been making of my ideas of 1927, has led me to propose a modification of the definition of the u wave. In 1927, I considered it as a solution with a singularity of the *linear* equations as assumed by wave mechanics for the ψ-wave. Various considerations, in particular a comparison with the theory of General Relativity of which I shall speak again, have made me think that the true propagation equation for the u wave might be *non-linear* as are those that are met with in Einstein's theory of gravitation—a non-linear equation which would take the approximate form of the linear equation of wave mechanics when the values of u would be small enough. If this point of view is correct, it might even be assumed that the u wave does not contain a moving singularity in the strict sense of this word, but simply a very small mobile region of singularity (of dimensions doubtless of the order of 10^{-13} cm.) within which the values of u would be large enough so that the linear approximation is no longer valid, although it would still be valid in all space outside this small region. Unfortunately, this change in viewpoint does not help solve the mathematical problems which were raised, for if the study of solutions with singularities for linear equations is often difficult, that of solutions for non-linear equations is even more difficult.

Let us return to 1927. Lorentz had asked me in the spring to prepare a report on wave mechanics for the 5th Solvay Congress of Physics, which was to be held at Brussels in the following October. Conscious of the difficulties that I would have to surmount in order to expound my ideas on the double solution, which were scarcely satisfactory from the point of view of mathematical rigor, I resolved to adopt a simpler point of view, the possibility for which I had pointed out at the end of my article in the *Journal de Physique*. Since, in my ideas of that time, the motion of a corpuscle was defined by the gradient of the phase which is common to the u and ψ solutions, everything takes place as if the corpuscle were "guided" by the continuous ψ-wave. It seemed to me that one could therefore take the following point of view: postulate the existence of the corpuscle as an independent reality, and assume that the corpuscle is guided in its motion by the ψ-wave according to the formula that the velocity is proportional to the gradient of the phase of ψ. This way of presenting matters I had designated by the expressive name of the "theory of the pilot-wave," and it was this that I developed in my report and which appears in the summary of the 5th Solvay Congress. I did not perceive at that time that, by adopting this kind of line of reasoning, I had greatly weakened my position. In fact, if the hypothesis of the double solution is difficult to justify mathematically, it would be capable, in the case of its success, of offering a very profound view of the constitution of matter and of the wave-corpuscle duality, and even perhaps as we shall see, permit a *rapprochement* of quantum and relativity conceptions. The simplified theory of the pilot-wave, although it is in some ways a kind of consequence of the theory of the double solution, has none of these advantages. Since the statistical and purely fictitious character of the ψ-wave is something well established and seemingly admitted by all, the theory of the pilot-wave ends in the inacceptable result of making a determination of the motion of the corpuscle by a quantity, the continuous ψ-wave. This ψ-wave has no real physical significance in itself, but depends on the state of the knowledge of the person using it and must, therefore, change abruptly when new information modifies this knowledge. If the conceptions that I stated in 1927 are some day to rise again from their ashes, they can only do so in the subtle form of the double solution and not in the crippled and inacceptable form of the pilot-wave.

At the Solvay Congress of 1927, my exposition of the pilot-wave received a poor reception. To my concepts, Pauli made some serious objections to which I could half-see a possible answer, but without being able to make it completely precise. Schrödinger, not believing in the existence of corpuscles, could not go along with me. Bohr, Heisenberg, Born, Pauli, Dirac . . . were developing the purely probabilistic interpretation which is now the "orthodox" interpretation. Lorentz, the president of the Congress, could not subscribe to such an interpretation, and emphatically reaffirmed his conviction that theoretical physics must remain deterministic and continue to use clear images in the classical framework of space and time. Einstein criticized the probability interpretation and directed against it some disturbing objections: he encouraged me to continue on the way I had started, but without giving a clear approval of my endeavor.

I returned to Paris much troubled by these discussions and, in meditating on this subject, I came to the conclusion that, for the reason I stated above and for several other ones, the theory of the pilot-wave was untenable. Not daring to return to the double solution because of its mathematical difficulties, I became disheartened and aligned myself with the purely probabilistic interpretation of Bohr and Heisenberg.

For twenty-five years, almost all physicists have aligned themselves with the purely probabilistic interpretation of Bohr and Heisenberg. There are however several notable exceptions, scientists as famous as Einstein and Schrödinger who have always refused to accept it and have attacked it with damaging objections. During the Solvay Congress of 1927, Einstein had developed the following objection: Let us consider a flat screen pierced by a hole which a corpuscle with its associated wave will strike perpendicularly. The ψ-wave is diffracted during its passage through the hole, and takes, behind the screen, the form of a divergent spherical wave. If a hemispherical film is placed behind the screen, the localization of the corpuscle at a point P on this hemisphere can be registered by a photographic impression. Wave mechanics tells us (and all agree on this point) that the probability of a localization at P is given by the square of the amplitude of the ψ-wave. If there exists at each instant a localization of the corpuscle, permitting us to define a trajectory (with the help of hidden variables), we can easily see that our ignorance of the trajectory of the corpuscle would permit us only to define a probability that the trajectory would pass through such or such a point of the film: the fact that the corpuscle produces a photographic action at P tells us that the trajectory of this corpuscle did pass through P and, as soon as we have this information, the probability of the trajectory passing through other points of the film vanishes. This is a straightforward explanation, but it is not at all that which a purely probabilistic interpretation would give. According to the latter, before the photographic impression, the corpuscle is potentially present in all points of the region behind the screen with a probability equal to the square of the amplitude of the ψ-wave. As soon as a photographic impression is made at P, the corpuscle is localized, is condensed one might say, at the point P and *instantly* the probability of its being at any other point of the film falls to zero. Now, Einstein said, such an interpretation is incompatible with all our ideas about space and time (even in their relativistic space-time form) and with the idea of a propagation of physical actions in space with a finite velocity. And it is insufficient to say that our concepts of space and time as drawn from macroscopic experience could be in error on the atomic level: actually, the film does have macroscopic dimensions (it might have a surface of a square meter) and here then it would be a question of the insufficiency of our notions of space and time *even* on the macroscopic level, and this seems really difficult to believe. To this objection of Einstein, to which as far as I know no one has given a satisfactory answer, other facts have subsequently been added by Schrödinger and again by Einstein relating to phenomena of interaction. I can not outline these arguments here, I shall only say that like Einstein's objection of 1927, they lead to paradoxical conclusions, in particular to a doubt, even on the macroscopic level, of our former notions of space and time.

That then was the situation, just about stabilized for the last quarter century, when there appeared, some months ago, the article of Mr. Bohm of which I spoke at the beginning. This article contains nothing essentially new since he has only taken up again the theory of the pilot-wave that I had expounded at the Solvay Congress, a theory which, by making appeal only to the probability wave ψ and not to the u wave with a singularity introduced by the hypothesis of the double solution, had always seemed to me to run into insurmountable difficulties. Nevertheless, in addition to the merit of having redirected attention to these questions, Mr. Bohm also had that of making a certain number of interesting remarks and, in particular, of making an analysis of the processes of measurement as seen from the point of view of the pilot-wave which would seem to permit removing the objections which had been directed against my ideas by Pauli in 1927. As soon as I had knowledge of the article of Mr. Bohm and the ideas of M. Vigier, concerning which I shall speak in a moment, I made a résumé of my ideas on this subject in two notes in the *Comptes Rendus* of the Academy of Sciences, which appeared in the September 51 and January 52 issues. One of the points which attracted my attention is the following: the argument of von Neumann [1]

[1] See footnote at the end of Chapter 10, Sect. 4. Tr.

claims to prohibit any interpretation of the probability distributions of wave mechanics by a causal theory of hidden parameters: now the theories of the double solution and of the pilot-wave, if they can not be considered as being proved, do however *exist* and we might wonder how their existence can be reconciled with von Neumann's theorem. This remark has led me to a reëxamination of the demonstration of this theorem, and I have now seen that this demonstration rests essentially on the following postulate: all probability distributions assumed by wave mechanics have a physical existence *even before an experiment has been made which brings one of the distributions into play*. So the probability distributions deduced from knowledge of the wave, and relative to the position and state of motion, would exist before experiments with measurements which might permit us to know exactly the position and state of motion. Now it could very easily be assumed, on the contrary (and this is even quite in agreement with the essential role that all quantum physicists today attribute to acts of measurement) that these probability distributions or at least certain of them, can be created by the execution of a measurement and can exist only when a measurement has been made, but before one has had any knowledge of the result. In the theory of the double solution and the pilot-wave (which in this respect are not distinct) it is assumed that the probability distribution relative to position as given by the square of the amplitude of the continuous ψ-wave exists before any measurement, but other probability distributions (for example, the one relative to momenta) would be created by the measurement: the postulate on which von Neumann's argument rests is therefore not applicable to the former and this destroys the conclusion of the argument. The purely probabilistic interpretation assumes an absolute equivalence of all the probability distributions, and that is why von Neumann has assumed this equivalence as a postulate, but in doing so he has simply shown that if we assume the conceptions lying at the base of the purely probabilistic interpretation we can no longer escape from this interpretation. Hence there is a sort of vicious circle, and the theorem of von Neumann does not seem to have any longer the import which even I had attributed to it in recent years.

After this beginning by Mr. Bohm, M. Vigier, who works at the Henri Poincaré Institute, had the most interesting idea of establishing a parallel between the theory of the double solution and a theorem proven by Einstein (likewise in 1927, but quite independently of my researches, for I was then doing work on quanta without busying myself with general relativity, while Einstein concentrated his attention on general relativity without busying himself with quanta). In order to comprehend the interest in this parallel, it is necessary to understand that theoretical physicists are at the present time divided into two camps which seem irreconcilable. Einstein and his pupils form a small group who seek to further the ideas of relativity by striving to extend the concepts of general relativity, while the large majority of theorists, attracted by an interest in atomic problems, have worked to further the progress of quantum physics without occupying themselves in any way with the ideas of general relativity. To be sure, wave mechanics has taken the concepts of special relativity into account, and has tried to incorporate them: Dirac's theory of electron spin, and more recently the splendid theories of Tomonaga, Schwinger, Feynmann and Dyson have used the ideas of relativistic covariance.[1] But it has always been special relativity which has been involved: now we know that special relativity is not sufficient in itself and that it must be generalized as Einstein did in 1916. It is therefore paradoxical that the two major theories of contemporary physics, the theory of general relativity and that of quanta, are today without any contact and mutually ignore each other. It is most necessary

[1] These theories are directed to a rigorous and complete relativistic theory of systems of particles which is necessary to solve the problem of extending wave mechanics to systems as discussed in Chapter 12 at the end of section 1. Tr.

that one day someone succeeds in effecting a synthesis of them.

After having worked out the major outlines of general relativity, Einstein was preoccupied with the manner in which the atomic structure of matter could be represented by singularities in the gravitational field. Then too, he was also preoccupied with the following point: in general relativity, it is assumed that the motion of a body is represented in curved space-time by a geodesic of this space-time, and this postulate had permitted Einstein to re-derive the formulae for the motion of the planets around the sun, which, in addition, explain the secular advance of the perihelion of Mercury. But if we would like to define the elementary particles of matter by the existence of singularities in the gravitational field, it must be possible to prove, *starting from the equations of the gravitational field alone*, that the motion of the singularities takes place along the geodesics of space-time, without having to introduce this result as an independent postulate. For a long time, this question had preoccupied Einstein who in 1927, in a work in collaboration with Grommer, succeeded in proving the theorem he had in mind. This proof has subsequently been taken up and extended in various ways by Einstein himself and his co-workers, Infeld and Hoffmann. It is certain that the proof of Einstein's theorem presents a certain analogy with that which I had given in 1927 to prove that a corpuscle must always have its velocity directed along the gradient of the phase of the ψ-wave, of which it constitutes a singularity. M. Vigier is vigorously investigating attempts to make this analogy precise by seeking to introduce the u wave functions into the definition of the metric of space-time. Although these attempts are perhaps not yet fully convincing, it is certain that the direction in which he is going is most interesting for it might lead to a unification of the ideas of general relativity and wave mechanics. By representing material corpuscles (and similarly photons) as singularities in the metric of space-time surrounded by an undulatory field of which they would be a part and whose definition would bring in Planck's constant, one would succeed in uniting Einstein's conceptions about particles and those of my theory of the double solution. The future will say whether this grand synthesis of Relativity and Quanta is really possible.

To me one thing is certain, which is that in such a synthesis, we shall have to re-derive and justify all the results, all the methods of calculation used by wave mechanics in its current interpretation, including the impossibility of predicting, in general, the results of a microphysical measurement, the Heisenberg uncertainties, the quantization of atomic systems, etc. But then, one will say, why modify the current interpretation, if it is sufficient to account for all the observable phenomena, why introduce all these useless complications of the double solution, of solutions with singularities, etc., thus risking meeting up with new impasses? To this, it can be answered, first of all, that a return to clear Cartesian conceptions, respecting the validity of space and time, would certainly satisfy many minds and would permit us not only to answer the troublesome objections of Einstein and Schrödinger, but also to avoid certain strange consequences of the present-day interpretation. Actually, this interpretation, by seeking to describe quantum phenomena solely by means of the continuous ψ-function whose statistical character is certain, logically ends in a kind of "subjectivism" akin to idealism in its philosophical meaning, and it tends to deny the existence of a physical reality independent of observation. Now a physicist instinctively remains a "realist," and he has several good reasons for this: subjective interpretations always give him a feeling of uneasiness and I believe that in the end he would be happy to be free of them.

But it might also be thought, along with Mr. Bohm, that if the present-day interpretation is sufficient for the prediction of phenomena on the atomic scale (10^{-8} to 10^{-10} cm.) it might not be the same on the nuclear scale (10^{-13} cm.) where the singular zones of the various corpuscles might overlap and could

no longer be considered isolated. It must be confessed that a
the present moment the theory of nuclear phenomena, in par
ticular of the forces which maintain the stability of the nu
cleus, is in a very unsatisfactory state. Besides, a theory of ma
terial corpuscles is at this moment sorely needed by us becaus
we are discovering almost every month a new kind of meson.[1] I
seems that physics has urgent need of being able to define
structure for these particles, in particular to be able to introduc
a "radius" for the electron, as in the old Lorentz theory. Now i
has found itself very much hindered in doing this by the exclusiv
use of the statistical ψ-wave to describe the particles, for it pro
hibits the use of any structural image for these particles. It i
permissible to believe that a change in viewpoint embodying
return to spatio-temporal images will help this situation: obviousl
this is only a hope, a blank check as Pauli would say, but th
possibility, we think, should not *a priori* be completely exclude
and it is necessary to avoid the danger that too great a faith i
the purely probabilistic interpretation of quantum physics doe
not finally make it sterile.

The question which finally must be answered is knowin
(Einstein has often emphasized this point) whether the pres
ent interpretation, which uses solely the ψ-wave with its sta
tistical character is a "complete" description of reality—i
which case it would be necessary to assume indeterminism an
the impossibility of representing reality on the atomic leve
in a precise way in the framework of space and time—or if, o
the contrary, this interpretation is incomplete and hides be

---

[1] The existence of the meson, postulated for theoretical reasons by Yukaw
in 1935, was first confirmed by laboratory experiments in 1948. But toda
the existence of five kinds of mesons is considered to be definitely establishe
(+ mu mesons, — mu mesons, + pi mesons, — pi mesons, and neutral p
mesons) and the existence of four additional kinds is considered probabl
(+ kappa mesons, — kappa mesons, + tau mesons, — tau mesons). For
discussion of the properties of these and of the other elementary particle
of matter, see article by Robert E. Marshak in *Scientific American*, Janu
ary, 1952, pp. 22-27. Tr.

hind itself, as the older statistical theories of classical physics do—a perfectly determinate reality, describable in the framework of space and time by variables which would be hidden from us, i.e., which would escape experimental determination by us. If this second hypothesis is to prove fruitful, it is, it seems to me, in the form of a theory of a double solution, more or less modified and doubtless brought into agreement with general relativity, that it will have to be made explicit. But I am not ignoring (and this recent review of the whole question has convinced me more of this) the tremendous difficulties, perhaps even insurmountable, such an attempt is going to run into, and what difficult mathematical justifications would be necessary to establish it on solid ground. If this enterprise should prove beyond fulfillment, it then would be necessary to return to the purely probabilistic interpretation, but at the present time a new examination of the question does not seem superfluous to me.

Doubtless, after having seen me abandon my original endeavors in this direction and then expound the interpretation of Bohr and Heisenberg in all my writings during 25 years, some will perhaps accuse me of inconsistency at seeing me express some new doubts on this subject, and will ask me if my first attitude was not after all the right one. To this, if I may joke, I can answer with Voltaire: "A stupid man is one who doesn't change." But a more serious response is possible. The history of science shows that the progress of science has constantly been hampered by the tyrannical influence of certain conceptions that finally came to be considered as dogma. For this reason, it is proper to submit periodically to a very searching examination, principles that we have come to assume without any more discussion. For a quarter of a century the purely probabilistic interpretation of wave mechanics has certainly been of service to physicists because it has kept them from being overwhelmed by the study of very arduous problems which are as difficult to solve as are those that the conception of the double solution poses, and thus has permitted them to advance steadily in the direction of applications, which have been numerous and fruitful. But today the heuristic power of wave mechanics, as it is taught, seems in large measure weakened. Everyone recognizes this and the partisans of the purely probabilistic interpretation themselves are searching, without much success it would seem, to introduce new concepts which are even more abstract and further removed from classical images, such as s-matrices, minimal length, non-linear fields. Without denying the interest of these endeavors, it might be asked if it is not rather toward a return to the clarity of spatio-temporal representations that we must direct ourselves. In any case it is certainly of use to take up again the very difficult problem of an interpretation of wave mechanics in order to see if what is now orthodox is really the only one that can be adopted.

Chapter 11 The Spin of the Electron

1. The Fine Structures and the Magnetic Anomalies

We have developed the principles of the wave mechanics of the electron. We must now show why, despite its success, this mechanics in its original form still seemed insufficient and still had to undergo an important re-working. The reason for this is that the wave mechanics of the electron in its original form did not succeed in interpreting certain facts of a spectroscopic or magnetic nature which had been known for several years and for which the old quantum theory could give no better account.

The first category of these difficult to explain facts belongs to spectroscopy. We know that the old quantum theory and, afterwards, the new mechanics had succeeded in correctly predicting the existence of a great number of the spectral lines.

But the tables of spectral terms obtained by these theories were shown, upon examination, to be insufficient to account for the actual complexity of the spectra. In other words, there are in optical and Röntgen spectra structures which remained without interpretation. We have seen that Sommerfeld, working with the help of relativity ideas within the framework of the old quantum theory, had succeeded in accounting for the fine structure of the hydrogen spectrum and of the X-ray spectra, in a way which at first seemed very satisfactory, but we have also emphasized (Chapter 6, at the end of section 3) that a more attentive study had not completely confirmed this favorable impression: Sommerfeld's theory, as we have said, correctly predicts the doublets of the Balmer series and the X-ray series, but it does not put them where they really are. The apparent success of Sommerfeld could not be thought of as being entirely fortuitous, but it was felt that an important element must be lacking in his theory. The situation, far from being bettered by the development of wave mechanics, was on the contrary aggravated. Actually, to be able to translate Sommerfeld's attempt into wave mechanics, it was necessary to introduce relativity into it. Now, as we have seen, a relativity wave equation was easily found which, besides being of second order with respect to time, appeared to be the natural relativistic generalization of Schrödinger's equation. It seemed that it should be enough to apply the new method of quantization to this equation, i.e., to find its proper values, in order to obtain Sommerfeld's formulae once again. The outcome of this calculation was disappointing: a formula was found that was of a somewhat analogous form to that of Sommerfeld's but nevertheless different, and this formula did not correspond any better to the experimental facts that were to be explained. The failure accordingly was complete: wave mechanics had not introduced the new element which was needed and whose nature was already known at this time through the work of Uhlenbeck and Goudsmit, concerning which we shall speak later on.

But, besides questions relative to the doublets of Sommerfeld, other difficulties also arose concerning the fine structures. Thus, in X-ray spectra, Sommerfeld's theory did predict very well some of the fine structures which actually exist, but the structure of the series is a great deal more complex than the formulae of this theory would indicate. To give an example of this, in the X-ray spectra of the elements there are always three L series whose lines overlap in the scale of frequencies. Sommerfeld's theory permits two L series to be predicted and two only: it completely ignores the third. To permit a classification of the unforeseen spectral lines, Sommerfeld later introduced, alongside the two quantum numbers contained in his theory, a third quantum number to which he gave the rather unjustified name of "inner quantum number." The introduction of this third quantum number was out and out empirical and attempts at a theoretical interpretation which were proposed for it at that time had to be abandoned. Further, wave mechanics was no more fortunate and had no success in interpreting the existence of the supernumerary series and the inner quantum number. Here again, there was felt the necessity of introducing the new element of which I have spoken above.

Let us now turn to the second category of phenomena lacking explanation in the old quantum theory: the magnetic anomalies. We have already pointed out the existence of the anomalous Zeeman effects that neither the original electron theory of Lorentz, nor the old quantum theory, nor wave mechanics had succeeded in explaining. The reason for this common failure is found in the fact that one and the same postulate is put at the base of the interpretation of the Zeeman effect in the three successive doctrines. This postulate is that the magnetic moments which atoms can possess all have as their sole origin the orbital motions of the intra-atomic electrons. This point assumed, it necessarily follows that the total moment of momentum of an atom always takes a certain fixed ratio to its total magnetic moment, depending solely on the ratio of the electric charge of the electron to its mass. This result, common to the classical electron theory, to the old quantum theory and to wave mechanics in its original form, leads in each of these three disciplines to the consequence that all the Zeeman effects should be of the normal type originally predicted by Lorentz and discovered by Zeeman. The existence of anomalous Zeeman effects indicates, just as the existence of the spectroscopic facts of which we spoke above, the necessity of introducing a new element, and shows that this element must have something to do with magnetism.

The experimental study of the anomalous Zeeman effects had, moreover, been pursued without let-up since their memorable discovery by Zeeman, and empirical laws for them were very well known. We can not discuss these empirical laws here and shall limit ourselves to saying that Landé had succeeded in summarizing a great number of them by introducing a certain factor into the formulae of the old quantum theory, Landé's *g* factor, for which the correct interpretation remained doubtful. The whole body of this research on the anomalous Zeeman effect doubtlessly paved the way for a complete theory of the phenomenon since we thus knew in advance the exact mathematical form of the laws that had to be accounted for.

But the anomalous Zeeman effects were not the only phenomena of a magnetic kind which remained unexplained: there were also gyro-magnetic anomalies. The hypothesis that atomic magnetism has its origin in the orbital motion of electrons in the atom implies that, if a cylindrical iron bar suspended from a point on its axis is magnetized, this bar should begin to rotate about this axis, and conversely, if the bar is set in rotation about this axis, a magnetic moment should be created: besides, the ratio of the moment of momentum due to the motion of rotation to the magnetic moment, in either case, ought to be equal to the above mentioned constant which depends on the characteristics of the electron. Experiments were made to verify quantitatively this prediction of the theory (Einstein and de

Haas, Barnett). The two inverse phenomena exist: there is rotation of a magnetized bar and magnetization of a rotating bar. But the ratio of the magnetic moment to the kinetic moment is found to have about twice the predicted value. This unexpected result clearly indicated in which direction the new element to be introduced must be sought. It became apparent that all the magnetism of the atom does not have its origin in the orbital motion of electrons, and that there exists magnetic moments and kinetic moments whose ratio does not always have the values assumed up till then. By following this lead, Uhlenbeck and Goudsmit came to the important idea of the existence of an intrinsic rotation and intrinsic magnetism for the electron.

2. *The Hypothesis of Uhlenbeck and Goudsmit*

In a paper of great import, Uhlenbeck and Goudsmit proposed, in 1925, to consider the electron as possessing not only an electrical charge, but also a magnetic moment and a moment of rotation. It is very easy to get a classical picture of such a magnetic, rotating electron: it is sufficient to compare the electron to a little sphere full of negative electricity and turning about one of its diameters. Uhlenbeck and Goudsmit made their hypothesis precise by assuming that the ratio of the intrinsic magnetic moment of an electron to its intrinsic kinetic moment had a value twice the normal classical value. This hypothesis was suggested to them by the results of the gyro-magnetic experiments. Besides, it could be justified by means of the classical model of the electrified sphere in rotation, but this justification, by reason of the difficulties that exist from a quantum point of view in adopting this classical model, could not be considered very convincing. Nevertheless, as we shall see, the hypothesis of Uhlenbeck and Goudsmit was found to be remarkably well confirmed through its implications: the element missing in all previous theories had been found!

We should like to make the quantitative side of this new

hypothesis precise. In quantum theory, atomic electrons in their quantized states possess an orbital kinetic moment whose value is always a whole multiple of Planck's constant divided by 2π: this is the result of quantization itself. They also possess on orbital magnetic moment whose value is a whole multiple of a fundamental quantity called "the magneton of Bohr," which plays the role of a veritable atom of magnetism and whose role today is essential in all the general theories about magnetic phenomena. A famous experiment due to Stern and Gerlach, which permits the magnetic moment of an atom to be measured, has definitely confirmed the physical existence of Bohr's magneton. The quotient of Bohr's magneton by the quantum unit of kinetic moment, $h/2\pi$, furthermore has the classical value of which we have spoken in several preceding instances. Uhlenbeck and Goudsmit have assigned to the electron an intrinsic kinetic moment equal to one-half the quantum unit $h/2\pi$. Thus the ratio of the two moments is found to be equal to just double the classical value. They have used to designate the intrinsic rotation of the electron and the corresponding kinetic moment the English word "spin" which has since been favorably received and is used by all physicists.

At the time when these two Dutch physicists had their remarkable idea of introducing the spin of the electron, the new mechanics was just dawning. It is therefore most understandable that the new hypothesis should first be developed within the framework of the old quantum theory. First Uhlenbeck and Goudsmit, and then other physicists among whom we might mention Thomas and Frenkel, returned to the theory of the fine structure and the Zeeman effect by introducing into it the new properties which had just been ascribed to the electron. The results were most satisfactory, and clearly indicated that we were on the right road. The few difficulties that persisted were evidently due to the use of the old quantum methods and were destined to disappear as soon as the spin of the electron could be introduced into wave mechanics. This introduction was not effected without some difficulty, but finally Dirac, guided by an important work of Pauli, succeeded in bringing this about in a most interesting manner which has opened up all sorts of new perspectives. In order to be better prepared to approach the study of Dirac's theory, we must first say something about the preliminary work of Pauli.

3. *Pauli's Theory*

THE SPIN of the electron presents a certain analogy with the property of a photon which we call "polarization of light." It defines, in effect, a certain asymmetry, a certain lack of isotropy, in the electron. Surely, there is no complete identity, for the spin has a direction and a sense, while polarization, by reason of the vibration of the light vector, defines a direction, but not a sense in this direction. Nevertheless, it seems probable that if it is desired to introduce the spin into wave mechanics, it will be necessary to take our inspiration from the way in which polarization is compatible with the existence of a photon in the dualistic conception of light, for this inductive method is a continuation of the one that had led us to a theory of material waves by starting out from the known theory of light waves. This observation seems to have guided Pauli in the working out of his important researches on spin.

Let us therefore examine how we must seek a reconciliation of the polarization of light with the existence of the photon. Let us consider a beam of plane polarized light which falls on a Nicol prism. According to the classical theories of wave optics, everything takes place as if the presence of the Nicol prism had the effect of resolving the incident plane vibration along two perpendicular axes D and D' connected with the structure of the prism: the component along D is transmitted, the component along D' is stopped. If the Nicol is turned 90°, the axes D and D' could be considered as not having changed, but now it is the component along D' which alone is transmitted.

Therefore for any pair of axes D and D', perpendicular to the direction of propagation and at right angles to each other, the incident vibration can be resolved along D and D' and a Nicol prism, suitably oriented, will isolate one or the other of these components. Further, it will be the same situation, if the incident light, in place of being plane polarized, has any kind of polarization. Thus, to a given incident light, there corresponds an infinity of possible decompositions along two rectangular axes (normal to the direction of propagation) since these two axes can be oriented in their plane in an infinity of ways: to each of these decompositions there corresponds the possible isolation, by means of the Nicol prism, of two polarized beams at right angles to each other. Let us now take up the interpretation of the same phenomenon by assuming the existence of photons. A cloud of photons, associated with a wave of known polarization, enters a Nicol prism: a part of the photons pass through the prism and are found at the exit of the apparatus to be associated with a wave polarized in a direction D, the other photons being stopped. Now, according to the wave theory, the transmitted light energy is measured by the square of the amplitude, by the intensity of the component along D of the incident vibration and the light energy stopped by the prism is measured by the intensity of the perpendicular component. We are therefore forced to assume that the proportion of the incident photons whose polarization, after traversing the Nicol, is plane along D, is measured by the intensity of the component of the incident light vibration along D, while the proportion of the photons stopped by the Nicol is given by the intensity of the component at right angles. But nothing hinders us from supposing that the experiment be made with light of a very low intensity: the photons then arrive one after the other and, as happened for us before in the case of interference, we must substitute a probability point of view for a statistical one, and say that the probability that an incident photon will reveal itself, after traversing the Nicol, as being plane polarized along direction D is measured by the intensity of the component of the incident vibration along D. We can still say that, for each pair of rectangular axes D and D', there are two possibilities of plane polarization for the photon and that the respective probabilities of these two possibilities are given by the intensities of the two components of the incident vibration along D and D'. It is quite obvious that we have thus come to conceptions quite analogous to those which we have adopted to measure mechanical quantities. The Nicol might be considered as an arrangement allowing us to recognize whether an incident photon is polarized along D or D' and, if the state of the incident photon, represented by its associated wave is known, in general we will not be able to predict exactly the result of the measurement, but only assign a probability to the two possible hypotheses. Since there is an infinity of ways to choose the axes D and D', there is an infinity of possible plane polarizations contained potentially in the initial state of the photon, just as there are several values of the energy contained potentially in the state of a corpuscle whose associated wave is not monochromatic. Of course, it can happen in exceptional cases, that the result of the action of a Nicol on a photon can be exactly predicted: this will happen when the initial state of the photon is a pure case for the direction of polarization DD', in other words when the incident wave is plane polarized along D or D'. All that we have just said is retained without difficulty if, in place of considering a plane analyzer like the Nicol, a circular or elliptical analyzer is considered.

The result of all this is that, for a photon associated with any light wave, we can not ask the question: "What is the plane polarization of the photon?" This question has no meaning: it isn't capable of any reasonable answer. The only question that we can ask is the following: "What is the probability that an experiment (made with a plane analyzer) will permit us to attribute to a photon a plane polarization in a given direction D (normal to the direction of propagation)?" We have just seen how the wave theory furnishes us the answer and this answer rests essentially on the possibility of resolving the wave function into two components.

Pauli thought that, in order to introduce the spin of the electron into wave mechanics, it would be equally necessary to attribute two components to the ψ-wave, without moreover supposing that these two components would necessarily have the character of the rectangular components of a vector as in the case of light. Just as one can not in general speak of the plane polarization of a photon, so one would not be able to speak of the direction of spin of an electron. One might only wonder what the probability is for the spin of an electron to reveal itself as being in such a direction. But, as we have explained, the spin has a direction *and a sense;* further, it is supposed to have a value equal to a half-quantum unit of kinetic moment, or $h/4\pi$, Pauli therefore supposed that, for each direction D of space (which here is not constrained to be normal to the propagation, for ψ-waves are not transverse), the spin can have two values $\pm h/4\pi$, according to the sense it has in that direction. We must therefore ask ourselves: "What is the probability that an experiment would lead to attributing to the electron under consideration a spin directed along D with the value $+h/4\pi$?" And what is the probability that an experiment would lead to attributing to it a spin directed along D with a value $-h/4\pi$?" Pauli, by analogy with the polarization of light, assumed that for each given direction D, the wave is resolvable into two components whose intensities measure the respective probabilities of the two possible values $\pm h/4\pi$ for the spin in the direction D. Of course, if the direction D is changed, the resolution of the ψ-wave into two components will be made in a different way, just as for light the resolution of the vibration into two rectangular components is made in a different way depending on the system of rectangular axes considered. Pauli wrote out the two simultaneous differential equations which the two components of the ψ-wave must satisfy for a given direction D and he studied the manner in which the two components are transformed when the direction D is changed. He noted, by doing this, that the two components of the ψ-wave would not transform as vector components. Thus we have the first example in physics of a mathematical entity, viz., the ψ-wave of a corpuscle with spin, that does not fall into the general category of tensors, of which scalars and vectors, as we know, are but special cases. This mathematical entity of a quite new type has since been studied and it has been given the name of "half-vector" or "spinor."

We can not develop here the formalism of Pauli's theory: besides it has not had a wide application for it was very quickly replaced by that of Dirac. In addition, Pauli's theory is not relativistic; it therefore can not be used for a prediction of the fine structure in the sense previously indicated by Sommerfeld. But the conceptions of Pauli were of the greatest interest: they indicated how the introduction of spin into wave mechanics must be made by consideration of the probabilities of the two possible senses of the spin for a given direction and by the substitution of a ψ-function with several components for the single ψ-function. It remained for Dirac, in a brilliant endeavor, to succeed in perfecting this first rough draft.

4. The Theory of Dirac

DIRAC WAS certainly guided in his work by the ideas of Pauli, but he also had another guiding principle: that of creating a relativistic wave mechanics that was truly satisfactory. We have seen, in fact, that from the beginning of the definitive development of wave mechanics a relativistic wave mechanics had been proposed which would rest on a wave equation which was of second order with respect to time. Dirac submitted this attempt to close scrutiny and had concluded that it must be rejected. The principle objection that he directed against this attempt at a relativistic wave mechanics, is precisely that the equation of propagation in it *is* of the second order with respect to time. From this it follows, contrary to what happens in non-relativistic wave mechanics, that if any initial state represented by a certain initial form of the ψ-wave is given, the conservation of the total probability is not automatically assured. Now the automatic conservation of the total probability is an essential condition in order that the general principles of the new mechanics could be conserved. Dirac, pursuing this argument with powerful logic, came to the conclusion that the equation or equations of relativistic wave mechanics must necessarily be of first order with respect to time and that, consequently, by reason of the relativistic symmetry between space and time, they must equally be of first order with respect to the space coordinates. Then he showed by arguments on which we can not dwell, that, in relativistic wave mechanics, the wave function must have four components obeying a system of four simultaneous equations in partial derivatives which bodily replace the single equation of propagation of non-relativistic wave mechanics. Finally, Dirac looked for the way in which the equations of propagation and the components of the wave function are transformed when the system of coordinates is changed. He found, remarkably enough, that the equations are invariant for the Lorentz transformation, which at once makes his theory satisfactory from a relativity point of view. He gave the transformation formulae for the four components of the wave function, which are not those of a space-time vector, but belong, as will be better shown later on, to the new type of "spinorial" transformations, already encountered by Pauli.

But it is here that Dirac's theory holds a surprise: the equations of his theory, obtained by means of arguments of a strictly relativistic and quantum nature, where the spin hypothesis nowhere appears, do contain in themselves all the properties of a magnetic, rotating electron. In fact, it is easy to show that, by virtue of the new equations of propagation, the electron will behave as if it possessed an intrinsic magnetic moment equal to a Bohr magneton and an intrinsic kinetic moment equal to a half-quantum unit of kinetic moment. This possibility of making the spin spring from equations obtained quite independently of it is one of the most remarkable results in all contemporary theoretical physics, which has had its share of them.

We shall now show how Dirac's theory is based on Pauli's. In Dirac's theory, questions relative to spin must be put in the form that Pauli had indicated. We therefore must ask ourselves what is the probability for the spin to possess one or the other of those two possible values in such and such a direction D. To answer this question, it is necessary to find out first how the ψ-function is resolved into four components when the direction D is considered as the z-axis. The probability of one of the values $+h/4\pi$ will then be given by the sum of the intensities of the two components of even rank (the second and the fourth), while the probability of the other value will be given by the sum of the intensities of the components of odd rank (the first and third). Now a study of the solutions of Dirac's equations shows that, if the motion of a corpuscle is slow in relation to that of light, the first two components of the wave function are negligible in comparison to the last two. Said otherwise, when it is permissible to neglect the influence of relativity, it is sufficient to consider a wave function with two components, and then the intensity of one gives the probability of one of the possible values of the spin, while the intensity of the other component will give the probability of the second possible value. We have then returned to Pauli's theory exactly. This latter therefore appears to be only the non-relativistic Newtonian approximation of Dirac's theory. At the same time, it is understandable why there are four components of ψ in Dirac's theory in place of two in Pauli's: the existence of spin requires splitting the ψ-function into two components and the existence of relativity requires a further splitting of each of these two components, this second splitting not being needed in the Newtonian approximation.

We shall remark in passing that the whole probability interpretation of the new mechanics is very easily carried over into Dirac's theory at the price of a slight complication in the symbolism and we come to the applications of the new doctrine and to its successes. First of all, it permits an elucidation of the question of the fine structure and a definitive justification of the formulae of Sommerfeld, meanwhile correcting them. If, in fact, the quantization of the hydrogen atom is taken up again by means of Dirac's equations, it is seen that by reason of the appearance of the new element represented by the spin, a new quantum number, unknown in previous theories, is introduced and this coincides exactly with the "internal quantum number" empirically introduced several years before in the classification of the spectral terms as revealed by experiment. Thus, a formula for the fine structure is arrived at, having the same form as that of Sommerfeld, but where the new quantum number is substituted for the old azimuthal quantum number; and this substitution results in setting everything in order again, the predicted doublets now being found where experiment puts them. The same results extend to heavier atoms in so far as calculations can be made with simplifying hypotheses and the difficulties relative to the Röntgen doublets are cleared up. Thus it is proved that the essential idea of Sommerfeld, consisting in introducing relativity into the quantum theory to explain the fine structure was correct, but that the introduction of spin was equally necessary to obtain truly satisfactory results. The original success of Sommerfeld was not due to chance, but there was still lacking in his conceptions one essential element: spin.

Dirac's theory has also been quite fortunate in the interpretation of the magnetic anomalies. In dealing with the problem of the Zeeman effect, it found the existence of the anomalous effects which had so intrigued previous theorists. The reason for this success is simple to understand. In order to arrive at an explanation of the anomalous effects, it was essential to be able to attribute to the ratio of the magnetic moment to the kinetic moment of an atom a value different from the so-called "normal" one of which we have already spoken more than once. This normal value arises from the hypothesis that the magnetic moment of the atom derives solely from the orbital motion of its electrons. By attributing to an electron, in conformity with the hypothesis of Uhlenbeck and Goudsmit, a proper magnetic moment which bears to its proper kinetic motion a different ratio (double) from the normal ratio, Dirac's theory succeeds in escaping from the circle of normal Zeeman effects and in predicting the anomalous effects. But the success is not only qualitative, it is quantitative. A calculation, actually, permits us to justify the formulae of Landé and to predict values of the coefficient g introduced a little empirically by this author in a description of the anomalous effects.

The truly fine work of Dirac has therefore led to some remarkable results. It permits us to include within the realm of physical facts with a theoretical interpretation the whole body of spectroscopic or magnetic phenomena which through their resistance to all attempts at explanation had revealed the necessity of introducing spin. In a manner worthy of the greatest admiration, it realizes the union of the quantum point of view and the hypothesis of Uhlenbeck and Goudsmit. Evidently it can be asked how far it goes in reconciling and fusing quantum conceptions and relativity conceptions, the former implying essentially discontinuity, the latter being imbued with continuity. That is a difficult question which we do not wish to examine here. Our feeling is that the fusion of relativity conceptions and quantum conceptions is not completely realized in a satisfactory fashion by the theory of Dirac. But the edifice on the whole is admirable and constitutes the present day culmination of the wave mechanics of the electron.

Without stopping to examine other applications of Dirac's theory, for example, the problem of the scattering of radiation by matter (formula of Klein and Nishina), we should now like to develop a strange consequence of Dirac's equations which seemed, at the beginning, to constitute a weakness of the theory, but which finally turned out to be an advantage.

5. States of Negative Energy. The Positive Electron

THE EQUATIONS of Dirac's theory show a singular property of admitting of solutions corresponding to states of the associated corpuscle for which the energy would be negative. An electron in one of these states would have to have some strange properties: to increase its velocity, it would be necessary to withdraw some of its energy, to hold it back; to bring it to rest, it would on the contrary be necessary to give it some energy. Never had an electron, in an experiment, shown such an unexpected manner of behaving and there is good reason to believe that states of negative energy as permitted by Dirac's theory do not actually exist in nature. It might be said that in a sense the theory is too rich, at least in appearance.

The fact that Dirac's equations contain a possibility for states of negative energy doubtless springs from the relativistic character of these equations. In fact, even in the relativity dynamics of the electron as developed by Einstein at the beginning of Special Relativity, one also found the possibility of motions corresponding to a negative energy. But at that time, the difficulty was not very serious for Einstein's dynamics, in imitation of preceding theories, assumed that all physical processes were continuous. Now, since the proper mass of an electron possesses a finite value, the electron always has a finite internal energy, conforming to the relativity principle of the inertia of energy. Since this internal energy is not capable of annihilation, we cannot pass in a continuous fashion from states of positive energy to states of negative energy, and the then-current hypothesis therefore completely excluded such a transition. Hence it was sufficient to assume that at the beginning of time, all electrons were in states of positive energy in order to understand that it was always thus and thus would always be. The difficulty is a great deal more serious in Dirac's mechanics, for it is a quantum theory, admitting in essence discontinuities in physical phenomena. It can easily be seen that transitions between a state of positive energy and a state of negative energy would not only be possible, but even frequent. Klein has shown through an interesting example how an electron with positive energy, upon reaching a region of space where a rapidly variable field is acting, would be able to leave this region in a state of negative energy. Hence the fact that no electron with negative energy had ever been found in an experiment proved most awkward for Dirac's theory.

To remove this difficulty, Dirac had a most ingenious idea. Remarking that according to the exclusion principle of Pauli of which we shall speak in the next chapter, there could not be more than one electron in any state, he imagined that in the normal state of the universe, all the states of negative energy are occupied by electrons. From this it follows that the density of electrons in the negative state is everywhere uniform and Dirac supposed that this uniform density is unobservable. But there exists more electrons than are necessary to fill all the states of negative energy and the surplus constitutes the electrons of positive energy which reveal themselves to us in our experiments. In exceptional cases, an electron of negative energy, thanks to a transition brought on by an external agent, can pass into a state of positive energy: then there is a brusque appearance of an experimental electron and at the same time, formation of a "hole," a gap in the distribution of electrons with negative energy. Now Dirac proved that such a gap should be experimentally observable and should behave like a corpuscle which had the mass of an electron and with an equal and opposite charge: it must therefore show itself to us as an anti-electron, a positive electron. Furthermore, the gap accidentally formed will not be long in being filled by an electron with positive energy which will spontaneously undergo the transition accompanied by radiation, carrying it into the state of negative energy which was momentarily "empty." So, Dirac explained the non-observability of states of negative energy and at the same time, predicted the possible existence, even if exceptional and ephemeral, of positive electrons.

Dirac's hypothsis was, to be sure, ingenious, but it seemed offhand a little artificial. A greater number of physicists would perhaps have remained somewhat sceptical in this regard if experiments had not immediately come forward to prove the actual existence of positive electrons, whose general characteristics were just those as predicted by Dirac. In 1932, actually, the fine experiments of Anderson at first, and of Blackett and Occhialini later, showed that when atomic disintegrations of atoms were produced by cosmic rays, there appeared particles behaving exactly like positive electrons. Although it could not yet be asserted in an absolutely rigorous fashion that the mass of the new particles was equal to that of electrons and that their electric charge was equal and of opposite sign, later experiments have made this coincidence more and more probable. Further, positive electrons have shown a great tendency to be annihilated rapidly in contact with matter, with production of radiation. The experiments of Thibaud and Joliot would not seem to leave any doubt on this subject. The exceptional way in which positive electrons are produced and their capacity for annihilation which makes their life ephemeral, these are just the properties foreseen by Dirac, so that the situation turns out to be reversed: the existence of solutions of negative energy for Dirac's equations, far from putting them in jeopardy, shows that these equations include in themselves the existence and properties of positive electrons.

Nevertheless, we must recognize that Dirac's conception of "gaps" runs up against serious enough difficulties, notably concerning the electromagnetic properties of empty space. It seems probable to us that Dirac's theory is destined to be transformed in a way which will establish a greater symmetry between the two electrons and will cause the idea of gaps, together with the difficulties associated with it, to disappear. We shall take this up again in the next chapter. But withal, it remains no less true that the experimental discovery of positive electrons (today called positrons) constitutes a new and very remarkable confirmation of the ideas which lie at the base of Dirac's mechanics. The symmetry between the kinds of electrons which on close examination is stated by certain analytical peculiarities of Dirac's equations, certainly is of great importance and there is no doubt that it is destined to play an important role in the subsequent development of physical theories.

Chapter 12

The Wave Mechanics of Systems and Pauli's Principle

1. The Wave Mechanics of Systems of Corpuscles

UP UNTIL NOW we have been studying the new mechanics only for the case where a single corpuscle moves in a given field of force. Sometimes we have more or less implicitly assumed that analogous principles were valid in the case of a system, i.e., since physics assumes an essentially discontinuous character of the elementary physical entities, in the case of a group of corpuscles. We must now make precise how this wave mechanics of systems of corpuscles is established.

Let us remark at the outset that there truly is a "system" only if the corpuscles exert interactions on each other: without this, they can be considered separately and we again have the case of a single corpuscle. This remark was valid, of course, in the old mechanics as well as in the new.

Let us now recall how classical mechanics treated the problem of the motion of a system of corpuscles in interaction. For each of the corpuscles, it wrote down Newton's fundamental equation expressing the proportionality between the acceleration of a material point and the force which acts on it. Since, by hypothesis, there is an interaction, the force which acts on each of the corpuscles depends on the positions of all the others. The equations thus obtained therefore form a system of simultaneous differential equations. If the equations are written explicitly by using a system of rectangular Cartesian coordinates, their number is equal to three times the number of corpuscles since each corpuscle has three coordinates. The solution of this system of equations, when it is possible, furnishes an expression for each coordinate as a function of the time, i.e., it allows the position and motion of each corpuscle to be followed in the course of time. The solution of the equations that must be adopted is moreover completely determinate if at a certain initial instant the position and velocity of the corpuscles is given, in other words the instantaneous configuration and motion of the system. In this way mechanistic determinism proves to be valid in the classical dynamics of systems.

Without dwelling at any great length on the development of the classical mechanics of systems, we shall only recall that the equations of motion can be transformed and, under conditions frequently realized, put into the well-known forms of the equations of Lagrange and Hamilton. We shall refer the reader to what we have said about this subject in the first chapter. But, for these more abstract forms of the equations of motion, it is useful to adopt a new geometrical representation for the system. In place of our representing the system in a physical space of three dimensions by imagining at each instant the localization of each of the constituent corpuscles, we can, by joining together the coordinates of all the corpuscles, conceptually construct an abstract space having three times as many dimensions as there are corpuscles (this number of dimensions moreover being capable of reduction if there are relationships restricting the freedom of motion of the corpuscles). In this abstract space which is called the configuration space, each state of the system is represented by a point whose coordinates are equal to all the coordinates of the corpuscles of the system. The evolution of the system in the course of time will therefore be given by the displacement of this representative point in the configuration space. The whole mechanical problem then amounts to a calculation of the trajectory and the motion of this representative point, and the group of equations furnished by classical dynamics can be considered as the equations of motion of this point. We have thus reduced the study of the motions of multiple points in a physical, three-dimensional space, to the study of a single point in an abstract configuration space. Mechanistic determinism is then simply expressed by saying that the motion of the representative point is completely fixed if its initial position and velocity in the configuration space is known.

The use of configuration space becomes indispensable when one wishes to make use of Jacobi's theorem in the dynamics of systems. Interpreted in physical terms, this theory has as its essential goal the grouping of the possible motions in the problem at hand in such a way that, in each group, all of the possible motions correspond to all of the rays of a single wave propagation. It is obvious that if all the corpuscles in motion are represented in physical space it is impossible to establish such a correspondence because of the multiplicity of trajectories. On the other hand, it will be easy to establish this if configuration space is considered, for, in this space, a single trajectory of the representative point corresponds to each motion. Hence Jacobi's theory now permits us to classify the possible motions of the system, i.e., the possible motions of the representative point in the configuration space, so that the trajectories of the representative point belonging to one class represent in the configuration space the rays in a wave propagation in the sense of geometrical optics. Jacobi's equation which depends on all the coordinates of the corpuscles of the system, i.e., on all the coordinates of the configuration space, will be the equation of geometrical optics for this wave propagation in this space of multiple dimensions. The principle of Least Action will then appear to be equivalent to a Fermat principle. We have already explained all this in section 4 of our first chapter.

Since Jacobi's theory and the principle of Least Action open the royal road which leads from the old mechanics to wave mechanics, we might expect to see the latter develop within the framework of configuration space. This is just what has happened. By generalizing the procedure which had furnished him with the propagation equation for a corpuscle, Schrödinger succeeded in writing a propagation equation in configuration space for the ψ-wave associated with a system. This equation is so constructed that, if the approximation of geometrical optics is valid, we recover the Jacobi equation. But here the ψ-function depends, in addition to the variable time, on all the coordinates of all the corpuscles of the system, and its propagation takes place in configuration space. Here then the symbolic character of the ψ-wave is still more marked perhaps than in the case of a single particle. It might even seem quite strange that the motion of the system can not be dealt with in three-dimensional space, that, in order to do this, we must necessarily go through the intermediary of the abstract configuration space. In classical mechanics, the use of configuration space is often convenient, but always optional: all the corpuscles of the system can always be represented in physical space. The author of this book has long felt a certain uneasiness owing to the obligatory use of configuration space in wave mechanics: even today, he hopes that the laws of the wave mechanics of systems will be expressible in a less artificial form when we can replace our usual conceptions of physical space, of corpuscles, etc., by conceptions more adequate to reality.[1]

Be that as it may, the wave mechanics of systems is at the present time expressed by wave propagations in configuration space and we shall see that its methods have been crowned with success. The quantization of a system is effected by investigating for what value of the total energy of the system (which is equal to the frequency of the ψ-wave multiplied by h) there exists stationary ψ-waves in the configuration space, i.e., by seeking proper values of the propagation equations. Again, for these quantized systems, discontinuous spectra of proper values are found to which correspond proper functions forming a complete set. etc. Likewise the physical interpretation of wave mechanics is immediately generalized. The intensity of the ψ-wave will give at each point of the configuration space the probability that an experiment in the localization of the particles of a system will permit assigning to the system the configuration represented by the given point. Likewise, the partial intensity of the components of the spectral decomposition of the wave function into proper functions of the energy will give the probabilities that an experiment permitting an exact measurement of the energy of the system will assign as its value one or the other of the proper values of the Hamiltonian. In brief, all the principles of the probability interpretation are immediately carried over. We shall also note, in passing, that a center of gravity of the system can be defined and that some classical theorems of rational mechanics, such as that of Koenig, have their analogs in wave mechanics.

---

[1] In the case of systems containing particles of the same nature, this obligatory use of the abstract space of configuration can be avoided by the use of "super-quantization" or "second quantization." This method is based on the fact that in the evolution of any such system, the number of particles must always remain a whole number. Avoiding this abstract space would also be one of the achievements of the new theory of the double solution, which is discussed in Chapter 10, sect. 6. Tr.

The wave mechanics of systems, such as it has come from the works of Schrödinger, is not relativistic. It is the "waving," if it can be so expressed, of the Newtonian mechanics of systems and not of the Einsteinian mechanics of systems, and this for the good reason that this relativistic mechanics of systems has never been definitively set up. This impotence of relativity mechanics to deal rigorously with the motion of systems is tied to several causes and, in particular, to the fact that the relativity theory essentially rejects all instantaneous action at a distance. The relativistic wave mechanics of Dirac is applicable only to isolated corpuscles, placed in a known field of force: its generalization to the case of systems is a difficult problem which is far from a complete solution.

We shall deal in Section 4 with several of the finest applications of the wave mechanics of systems: but, before we do, we must study an important case where some circumstances quite characteristic of the new mechanics arise: the case of systems containing corpuscles of an identical nature.

2. *Systems Containing Particles of the Same Nature. Principle of Pauli*

THE MATTER that we are going to treat is dominated by an essential and quite new idea which arose in quantum theory when it was desirable to introduce the quantum of action into statistical mechanics. We shall explain in Section 5 how this introduction was carried out, but for the moment we shall limit ourselves to stating the idea to which it has led. It had always been assumed in atomic physics that two particles of the same nature, two electrons for example, were always identical. Nevertheless, this identity was not regarded as being so absolute as to prevent, at least in thought, distinguishing two corpuscles of the same nature: so, in statistical calculations for example, two states of the same system differing only by a transposition of the roles played by two particles of the same nature were regarded as being distinct. If, consequently, a system formed of electrons was being imagined, the collective state of the system where the first electron had an individual state *a* and the second an individual state *b* was regarded as different from the collective state of the system where the first electron had a state *b* and the second state *a*, the individual states of all the other electrons being the same in the two cases. The development of quantum statistics has led to a complete renunciation of the possibility of distinguishing two corpuscles of the same nature involved in a single system and to regarding two states of a system differing from each other only by a transposition of two corpuscles of the same nature as being identical and indistinguishable. We shall examine later what such a lack of individuality of the elementary corpuscles might signify. For the moment, we are dealing only with its consequences.

The transposition of corpuscles of the same nature has very important consequences in the wave mechanics of systems. Let us consider a system which contains corpuscles of the same nature and let ψ be one of the possible wave functions of the system. By definition, this wave function is called "symmetric with respect to two corpuscles" if, by a transposition of the coordinates of two corpuscles in its expression, its value doesn't change. It is, on the contrary, "antisymmetric with respect to two corpuscles" if, by a transposition of the coordinates of two corpuscles in its expression, it simply changes sign. It is essential to point out that, in general, a wave function is neither symmetric or antisymmetric. But the interchangeability of corpuscles of the same nature permits us to prove the following important theorem: "If a system contains corpuscles of the same nature, there always exists wave functions, some symmetric, some antisymmetric, with respect to all pairs of corpuscles of the same nature." Let us designate by the words "symmetric state of the system" a state whose wave function is symmetric, and by "an antisymmetric state of a system" a state whose wave function is antisymmetric. The fact that interaction potentials depend symmetrically on each pair of corpuscles then permits us to state another theorem no less important than the first: "It is not possible to bring about a transition bringing the system from a symmetric state to an antisymmetric one and vice versa." In other words, it is not possible that there be combinations, in the sense of Ritz, except between states of the same nature. From this it follows that symmetric states on one hand and antisymmetric ones on the other form two completely separated assemblages between which there exists no communication. Wave mechanics can therefore be reconciled with the existence of a principle which would assert that, for such and such a kind of corpuscle, only symmetric states or only antisymmetric ones are found in nature since, if at the beginning of time, only one sort of state was found to exist, it has always been and always will be the same: such a principle is not a consequence of wave mechanics which admits either kind of state, but it is compatible with it. We must now explain why Pauli was led to assuming the existence of such a principle, at least for electrons.

While studying the structure of an atom, we have pointed out (Chapter 4, sect. 4) the phenomenon of the saturation of energy levels and we have underscored its fundamental importance, for it is this which governs the whole development of the atomic structure along the series of elements and all the differences in chemical, optical or magnetic properties of these elements. We have also said that the manner in which the energy levels become successively saturated by the addition of new electrons had been empirically determined: it is summed up in a rule which was given by Stoner and which at first remained without a precise theoretical justification. Thanks to this rule of Stoner, the maximum number of electrons that each energy level of the atom can receive is thereby known. Seeking an interpretation of these facts, Pauli had the remarkable idea that the saturation of the levels had its origin in the impossibility for two electrons to have rigorously identical quantized states, i.e., defined by the same quantum numbers. Or otherwise stated, the presence of an electron in one quantum state would exclude the presence of any other electron in the same state: whence the name "exclusion principle" given to this new physical postulate. Translated into wave mechanics, Pauli's principle is expressed as follows: "for electrons, the only states realized in nature are the antisymmetric states." We have seen that such a statement is compatible with the new mechanics. In order to see that the two forms of the exclusion principle, as just given, are really identical, let us suppose that a system contains two electrons in the same individual state: if it is assumed, in agreement with the second statement, that the wave function is antisymmetric with respect to this pair of electrons, it must change sign if the role of the two electrons is interchanged, but since the two electrons are in identical individual states, this interchange can not thereby modify the wave function: thus, since the wave function must both change and not change its sign through the effect of this interchange, it necessarily is identically zero and this vanishing of the wave function in the new mechanics means that the state as imagined is non-existent. There therefore can not be two electrons in the same individual state and we see that the second statement leads us to the first: the converse is also easily proved.

Pauli's exclusion principle is therefore expressed analytically in wave mechanics by admitting as wave functions of a system containing electrons only wave functions antisymmetric with respect to all electron pairs. But, in the application of this principle, it really is necessary to remember that an electron possesses a spin so that its individual state is expressed as a function not only of its coordinates, but of the value of its spin. The wave functions admitted by Pauli's principle are those which are antisymmetric with respect to all the coordinates of space and the spin. We do not want to dwell here on this point whic is very important in the mathematical development of th theory.

Pauli's principle has the great merit of furnishing an interpretation for the saturation of the levels. It permits us to deduce correctly Stoner's rule by taking into account the fac that several different states, i.e., corresponding to differen combinations of quantum numbers, can have the same energ and hence belong to the same energy level: it is therefore sufficient to count up for each energy level how many of the different quantum states there are corresponding to this level i order to have, according to Pauli's principle, the maximum number of electrons corresponding to this level, since this maximun number is attained when each different quantum state i occupied. This calculation leads to Stoner's rule. We shall late see the fundamental importance of Pauli's principle in the application of the wave mechanics of systems: we shall also see hov it leads in the case of electrons to the Fermi-Dirac statistics.

Since, for electrons, antisymmetric states are the only possible ones, it might be asked what happens in this respect for th other elementary or complex particles of microphysics. I Pauli's principle applicable to them? Or rather, on the contrary, is it the symmetric states which are the only possibl ones? Or else are both states permissible? It seems certain tha this last alternative is never realized: it is always either antisymmetric or symmetric states which alone are realized in nature. The first case, as we know, is the one for electrons and als includes various atomic nuclei: there can not be more than on particle of this kind in each individual quantum state and th statistics is always, as we have seen, that of Fermi-Dirac. Th second case is represented by photons, α-particles and othe atomic nuclei: in this case there is no hindrance to the accumulation of any number of particles in one quantum stat since a symmetric function does not change through a transposition of two particles of the same nature: in the case of these particles with symmetric wave functions a particular statistics thereby results, the Bose-Einstein statistics, for which Planck's law is the expression for photons. In a general way, it seems that particles whose spin is an odd multiple of the unit of spin $h/4\pi$ obey Pauli's principle, while particles whose spin is zero or an even multiple of the unit of spin, follow the Bose-Einstein statistics. This is an important semi-empirical rule. These questions of spin and statistics play an important role in the study of band spectra and also in research on the constitution of atomic nuclei. Despite the importance of these considerations, we can not develop them here.

Pauli's exclusion principle expresses a very singular property of electrons and the other particles which are subject to it. Actually today it is almost impossible to understand how two identical particles can mutually debar each other from taking the same state. This is a kind of interaction quite different from those of classical physics and whose physical nature still completely escapes us. It would seem that this is a most important task, and moreover a very difficult one, for theoretical physics of the future to succeed in giving us an idea of the physical origin of the exclusion principle.

To show how far away from old conceptions we are in this domain, let us consider the case of a gas formed by particles of the same nature, obeying the Pauli principle, an electron gas for example. According to the exclusion principle, it is impossible for two electrons in the gas to possess the same state of rectilinear uniform motion, for here the quantized states are states of rectilinear uniform motion. With classical conceptions, this would mean that a particle situated at a point of the enclosure containing the gas would prohibit any other particle of the gas from taking the same state as it: and this is quite paradoxical since the enclosure containing the gas could be supposed as large as one wishes, and, consequently, the particles as remote as one wishes. But this paradox is intimately connected with the Uncertainty Relations of Heisenberg and vanishes if they are taken into account. In effect, the rectilinear and uniform motion of the particles correspond to well-determined energies for these particles: the Uncertainty Relations therefore forbid us to speak at the same time of states of motion of two particles and of their positions: the mere fact we speak of the energy states of the particles as being well-defined no longer permits us to speak of their separation since they are then in no way localized. This example shows us that a physical interpretation of the exclusion principle must necessarily be made completely outside the circle of classical images.

3. *Applications of the Wave Mechanics of Systems*

THE WAVE MECHANICS of systems, extended as required by Pauli's principle and by consideration of spin, has led to numerous, very brilliant successes. One of these was the interpretation of the helium spectrum. While the spectrum of ionized helium had been interpreted from the beginning by Bohr's theory (because ionized helium falls into the simplest category of a system with one electron) the spectrum of neutral helium had remained an enigma. The lines of neutral helium are actually divided into two altogether separate categories, corresponding to terms which, at least as a first approximation, do not combine. These two systems of completely independent lines had received the name of the spectrum of orthohelium and the spectrum of parhelium and for a long time it was thought that two different kinds of helium atoms existed, each emitting a different spectrum. But finally it was possible to recognize that there is not a distinct orthohelium and parhelium: the same helium atoms can, according to circumstances, emit the spectrum of ortho- or parhelium. Heisenberg, in a famous memoir, has given the key to this enigma. Since the two planetary electrons of the neutral helium atom follow Pauli's principle, the wave functions of this atom must be antisymmetric with respect to all of the coordinates and spins of the two electrons; but they can be so in two ways, either by being symmetric with respect to the coordinates and antisymmetric with respect to the spins, or by being antisymmetric with respect to the coordinates and symmetric with respect to the spins. There are thus two categories of wave functions and hence of spectral terms: furthermore since the spectral terms do not belong to the same category they can not combine, at least as a first approximation. It is then sufficient to identify one of the categories of terms with the terms of orthohelium and the other category with those of parhelium in order to obtain an altogether satisfactory explanation of the division of the two independent parts of the helium spectrum. By means of this interpretation, Heisenberg succeeded in accounting for several peculiarities of the orthohelium and parhelium spectra and, especially of this: while the lines of parhelium are simple, those of orthohelium are triple, forming triplets. The prediction of this small fact by Heisenberg's theory constitutes in itself a good verification of Pauli's principle, for this difference between the fine structure of the two series arises from the exclusion principle and without it, we would be led to different predictions which would not be in accord with experiment.

Another noteworthy application of the wave mechanics of systems is the theory of the hydrogen molecule, and more generally, the theory of homopolar molecules. Classical theory in a certain measure permitted us to comprehend the origin of the bond which united the atoms of a heteropolar molecule, i.e., of a molecule whose atoms have different electric attractions. In this case, indeed, it can be imagined that the various atoms of the molecules are changed into ions by borrowing or lending each other electrons: therefore, it can be thought that the stability of the molecular structure is assured by the play of Coulomb forces between the various ions of which it is made. But the case of homopolar molecules, for example the so important case of molecules formed by two atoms of the same nature, was most embarrassing to the older physics for there is no reason that atoms of the same electric affinity should change into ions of different signs and hence, it is no longer apparent what kind of force is going to be able to act as a bond between these neutral atoms, all those that can be imagined being a great deal too weak to play this role. Wave mechanics has permitted us, and this is not one of its lesser triumphs, to understand the nature of the homopolar binding thanks to the introduction of "exchange energies." Here is what this somewhat mysterious phrase means: when we closely examine by means of wave mechanics the evolution of a system containing identical particles, it is found that in the expression for the energy of the system, there appears, along with terms expressing the existence of known interactions between the particles, some terms of a new aspect connected with the possibility of transposing the identical particles. It is these terms which are designated by the name of "exchange energy." To them correspond forces of an altogether new type for which no vectorial representation in the classical manner is possible and which can have large values. These new actions are an inescapable consequence of the formalism of the new mechanics, but it appears quite impossible to make a physical representation of them in the old sense of this word. Once more, we find ourselves faced with a fact which transcends all classical conceptions and which shows us how mistaken is our usual method of localization of physical entities in a continuous space of three dimensions. It is very instructive to make the following remark: there is an exchange energy only if two identical particles have a non-zero probability of being found in one region of space. In other words, since the particles are generally not localized in wave mechanics, they have a certain probable density distribution: there is an exchange energy when two particles of the same kind have probable density distributions which overlap each other, and only in this case. This remark clarifies the relation existing between the exchange energy and the non-localization of particles in space.

Without dwelling further on these very interesting characteristics of exchange energy, we now should like to show how it explains the formation of homopolar molecules. The simplest case is that of the hydrogen molecule formed by two atoms, each containing one electron. When two hydrogen atoms, at first far apart, approach each other, they tend to form a mechanical system containing two electrons and an exchange energy between these two electrons appears. This exchange energy can be calculated by the methods of wave mechanics by taking into account the principle of Pauli and the existence of spin. That is what Heitler and London have done. The result of their calculations is as follows: if the spin of two electrons is in the same sense, the exchange energy corresponds to a repulsion between the atoms, and no molecule can be formed; if, on the contrary, the spins have a contrary sense, the exchange energy corresponds to an attraction between the atoms and becomes a repulsion if the atoms approach still closer, so that in this case there is a tendency to form a stable molecule. This theory accounts very well for the formation and properties of the hydrogen molecule. It can be expressed in essence by saying that the electrons of the two hydrogen atoms are capable of forming a pair with opposite spins and that such a pair, having a character of very marked stability, acts as the bond between the two atoms and keeps them united in a single molecule. Presented in this form, the explanation can be generalized for the formation of all bi-atomic molecules and even for molecules containing more than two atoms. Let us consider, for example, any bi-atomic molecule. The two atoms, which are capable of forming this molecule, contain a more or less large number of electrons: among these, a certain number form within the atom pairs of electrons of the same energy with opposite spins, but a few of them are not in this category. Now these electrons not involved in a pair, that have humorously been named "bachelor electrons," have a certain tendency, if occasion permits, to unite with electrons of another atom to form a pair. Calculations show, indeed, that in favorable cases the approach of two atoms will give rise to the formation of a molecule in which at least a part of the bachelor electrons of the two atoms will form pairs: it is the formation of these pairs which creates the molecular bond between the two atoms. This explanation obviously can be generalized for molecules formed by more than two atoms.

The interpretation of the formation of molecules through the constitution of electron pairs with opposite spins has permitted us to give an explanation of the notion of "valence," which is so fundamental in chemistry. In a general way, it can be said that an atom containing in its normal structure a certain number n of bachelor electrons will have a chemical valence equal to n. Such an atom will, indeed, be able to form a molecule with n hydrogen atoms since each of its n bachelor electrons can form a pair with the electron of the hydrogen atom: the given atom will therefore be n-valent or, at least, will have a maximum valence equal to n. It is therefore seen that the existence of chemical valence is connected with the exchange energy between electrons and this explains why no representation of valence forces by a vectorial scheme, usual in other cases, can give a truly satisfactory result. Besides, the fact that two electrons having formed a pair are somehow neutralized and can no longer contribute to any molecular union, accounts for the valence saturation, an absolutely incomprehensible fact as long as representation of valences by forces of the old type was sought for. It is therefore seen how useful and satisfactory to the intellect is the new theory of valence founded on wave mechanics.

But, if the new base for the theory of valence now seems certain, the detailed explanation of numerous chemical facts which are connected with this theory (multiple or directed valences, stereo-chemistry, free binding, etc.) is still hard work: it has already been most seriously begun, but this mathematical chemistry is a difficult science, and a lot still remains to be done before it is finished. Save in those simple cases, as that of the hydrogen molecule, the explicit calculation of proper values and proper functions is not possible: it is necessary at this time to be content with enumerating the proper values and classifying them according to the symmetry properties of the corresponding wave functions for which we can not write the expression. It is necessary at this time to use very general methods borrowed from the theory of groups. This theory, up to now little known to physicists, has thus become unavoidable in this branch of wave mechanics: with it we have moreover been led by rapid and elegant strokes to very good results of great generality. But since the physical theorists who know how to manage this difficult method have not always had the leisure to study all the numerous and complex basic facts of chemistry, a close collaboration must be established between them and the chemists in order to succeed in completing the results already obtained. As of today, it is in any case one of the most excellent claims to glory for this new mechanics to have been able to account for some of the most important laws of chemistry.

4. Quantum Statistics

THE METHODS of the classical statistical mechanics of Boltzmann and Gibbs, which had known such success in macroscopic physics, must of necessity have been affected by the development of the new mechanics. We can not explain here in detail the modification which the introduction of the quantum of action has produced in the very basis of statistical mechanics. We shall limit ourselves to giving an idea of it by considering the case of a perfect gas, using the images furnished by wave mechanics. In a perfect gas, the atoms, except for collisions, have states of rectilinear uniform motion: in classical statistical mechanics, these states of motion form a continuous series for all values and all orientations of velocity are equally possible. The methods of Boltzmann and Gibbs consist essentially in counting the possible distributions of the atoms of the gas between the states of motion for a given energy and seeking the most probable over-all distribution. At the time when the existence of the quantum of action was introduced by associating, as in wave mechanics, the propagation of a wave with the motion of an atom, the situation was modified for, the gas being enclosed in a fixed container, only the stationary waves in resonance with the dimensions of the container will be physically possible (according to the very conception of quantization in the new mechanics). It will therefore be necessary first to calculate the number of these stationary states, then to evaluate the possible distribution of atoms among the states for a given total energy. For a container having macroscopic dimensions, which is the only case practically realized, the stationary states form, because of the smallness of Planck's constant, a discontinuous, but extremely closely spaced, series. We might therefore believe that everything happens in practice as if the series were continuous, and hence that statistical mechanics is valid. This is correct in a large measure and explains the success of the old statistical methods, but, nevertheless, the introduction of the quantum of action has some verifiable repercussions on the macroscopic level. The principal one is that it permits a determination of the constant of entropy. In classical statistical mechanics, the constant of entropy is infinite, which seems most strange and, as we now know, springs from inadvertently neglecting the quantum of action, an indispensable element for the stability of the physical world. Some thought moreover to avoid this difficulty by saying that since the constant of entropy was arbitrary in thermodynamics, it little mattered if it were infinite! The quantum theory has permitted us to attribute a finite value to the entropy and to calculate it as a function of Planck's constant. It was then seen that the value of the constant of entropy entered effectively into the complete calculation of the equilibrium between a vapor and its condensate and this has permitted a quantitative verification of the value furnished by the quantum theory.

But, in order to develop completely the quantum form of statistical mechanics, it is necessary to evaluate the number of the various distributions of the atoms or other elements of the given system among the different possible quantum states, and as soon as the question has been asked, we must take into account the fact that the considerations developed in the section before last are going to play a very important role here. First, we have already seen that the identity of particles of the same nature obliges us to consider two distributions differing only in the transposition of two of these particles as being identical. This new way of counting distributions, which could have been used before in the old statistical mechanics since it is not a specifically quantum idea, has already given some results very different in principle from those of the Boltzmann-Gibbs statistics. But there are more: in making an enumeration of our distributions, it is going to be necessary to take account of the fact that our elements obey or do not obey Pauli's principle, i.e., if their wave functions are necessarily antisymmetric, there will be at most one of them in each state; if, on the contrary, they do not obey the Pauli principle, their wave functions then being, as we know, necessarily symmetric, nothing will limit the number of these elements being found in each possible state. In one case or the other, the enumeration will therefore be made quite differently. In the first case, we will have a statistics named after Fermi and Dirac, which could just as well be called the Pauli statistics, for it is potentially contained in the exclusion principle. In the second case, we will have a statistics named after Bose and Einstein, which is potentially contained in our original works on wave mechanics.

The two new statistics coalesce asymptotically with classical statistics if the value of h is made to approach zero, as could be predicted *a priori*. If two thermodynamics corresponding to these two statistics are developed, we obtain two slightly different thermodynamics which, too, will of course join up with classical thermodynamics for an infinitely small h. If the laws of a perfect gas are calculated in each of these cases, we obtain laws which depart in opposite directions from the classical laws: so, for example, in one case the gas is more compressible and in the other less compressible than the law of Mariotte-Gay-Lussac would indicate. Unfortunately, for gases in usual conditions, these departures, as we have stated before, are very small. For this reason it is impossible to detect them, all the more so since actual gases are not perfect and the departures with respect to the law of Mariotte-Gay-Lussac arising from other causes (interactions of molecules, finite volume they occupy, etc.) act to mask the departure which indicates the influence of the statistics. The new statistics do not therefore find their verification in the study of actual gases; but, fortunately, there exists for each of the two an important application which permits us to prove their exactness. On one hand, for the Bose-Einstein statistics it is the case of black body radiation, and on the other hand, for the Fermi-Dirac statistics the case of electrons in metals. We shall say a few words about each of these.

We have seen that photons do not obey the Pauli principle so that nothing prevents any number of photons from being in the same state. A gas formed of photons will therefore follow the Bose-Einstein statistics. Now the equilibrium radiation present in an isothermal enclosure is entirely comparable to a photon gas, with this difference however, that since the walls of the enclosure can absorb or emit radiation, the number of photons is not necessarily constant. By applying the Bose-Einstein statistics to the equilibrium radiation, taking account of the circumstance we have just given, Planck's law of spectral distribution is quite easily found. Since Planck's law is thoroughly verified by experiment, we have here obtained a remarkable confirmation of the Bose-Einstein statistics, and this confirmation is that much more convincing, since neither classical statistics, nor the Fermi-Dirac statistics could have led us to find the true spectral distribution of photons in equilibrium radiation.

The Fermi-Dirac statistics have found a remarkable verification in the electron theory of metals. The supporters of the old electron theory, in particular Drude and Lorentz, had sought an explanation of the properties of metals, especially their ability to conduct heat and electricity. They supposed that, in metals, the atoms are partially ionized, this ionization giving rise to a gas of free electrons within the metal. By applying the methods of statistical mechanics to this electron gas, they had succeeded in predicting satisfactorily a great many of the properties of metals. Nevertheless, in this theory, many difficulties still persisted: one of the most important was relative to the specific heat of metals, which by reason of the presence of the free electrons in the metal, should have been a great deal larger than it actually is. The development of the new statistics permitted Sommerfeld to solve a part of these difficulties. Since electrons are subject to the exclusion principle, they must follow the Fermi-Dirac statistics. Now a simple numerical calculation shows that the conditions in which the electrons in a metal are found are very different from those in which the atoms of an ordinary macroscopic gas are found: although for these latter the results furnished by the Fermi-Dirac statistics do not sensibly differ from the results furnished by classical statistics; for the electrons in a metal, on the contrary, the Fermi statistics do not give the same results at all as those of Boltzmann. This divergence springs from the extreme lightness of the electron with respect to material atoms. If the validity of quantum statistics is admitted, it is then necessary to rework completely the theories of Drude and Lorentz. It was Sommerfeld who first did this. In this way, he retained the correct results of the old theory, meanwhile perfecting them, and he solved a great many of the difficulties that it had raised. For example, he easily explained, by the very results of the Fermi-Dirac statistics, why the free electrons do not sensibly contribute to the specific heat of the metal and hence, why the latter has almost the same value as if there were no free electrons: so a great obstacle encountered by the old theory is removed. Numerous theorists, among whom we shall mention Leon Brouillon, Felix Bloch and Peierls, have been occupied with the way which was opened by the work of Sommerfeld, and have extended in various ways the results first obtained. Here is an entire, important branch of quantum physics which the size of this work unfortunately will not allow us to cover. We shall note that alongside these brilliant results, there still remain some shadows here: for example, the so curious and important phenomenon of superconductivity has not yet been satisfactorily accounted for.

Among other applications of quantum statistics, we shall limit ourselves to citing, without elaboration, the good use that Fermi has made of his statistics in an investigation of the properties of atoms by boldly considering each atom as a gas containing several electrons placed in the central field of the nucleus.

5. *The Limits of Individuality*

THE WAVE MECHANICS of systems containing particles of the same nature and their quantum statistics imply, as we have seen, a certain renunciation of the idea of individuality of the particles. It would seem a little excessive to us, however, that it is necessary to give up completely the idea of the individuality of particles. It seems to us that the possibility of individualizing particles is connected with the possibility of localizing them in different regions of space. This latter possibility always exists so that there is always a possibility of individualizing the particles by localizing them through experiment in different places in space. But the individuality of identical particles ceases to be capable of being "followed" when the possible probability density distributions of the particles overlap each other, for an exchange of particles then becomes possible: this refers to what we have said on the subject of exchange energy in the section before last. It is this latter case which is realized in most of the systems contemplated by wave mechanics, in particular in a gas where the particles are supposed to possess a well-determined energy, i.e., associated with a plane monochromatic wave (or almost so) which fills all the enclosure. We then understand why the non-individualization of particles can not arise in classical theories since it is connected with the possibility of two particles occupying, at least potentially, the same region of space, a possibility which is characteristic of the conceptions of the new mechanics.

If one will but reflect a little on certain remarks made in Sections 3 and 4, one will see that the non-individuality of particles, the exclusion principle and exchange energy are three intimately connected mysteries: all three derive from the impossibility of representing exactly elementary physical entities in a three-dimensional space continuum (or more generally in the four-dimensional space-time continuum). Perhaps some day, by escaping from this framework, we shall succeed in penetrating better the meaning, still quite obscure today, of these great guiding principles of the new physics.

From another point of view, it might be said that the physical notion of an individual is complementary, in the sense of Bohr, with the notion of system. The particle truly has a well-defined individuality only when it is isolated. As soon as it enters into an interaction with other particles, its individuality is diminished. Perhaps it has not been pointed out enough in classical theories that the notion of potential energy of a system implies a certain weakening of the individuality of the constituents of the system through the "pooling," in the form of the potential energy, of a part of the total energy. In the cases contemplated by the new mechanics, where particles of the same nature occupy, somehow simultaneously, the same region of space, the individuality of these particles is dissipated to the vanishing point. In going progressively from cases of isolated particles without interactions to the cases just cited, the notion of the individuality of the particles is seen to grow more and more dim as the individuality of the system more strongly asserts itself. It therefore seems that the individual and the system are somewhat complementary idealizations. This, perhaps, is an idea which merits a more thorough study.

Epilog — Concerning Several Questions Which Were Not Treated in This Book

1. Wave Mechanics and Light

We have seen how the fundamental ideas of wave mechanics were suggested by the dual nature of light. It is by extending to matter those conceptions to which we had come while reflecting on the association of photons and light waves that we succeeded in conceiving of and interpreting the association of material corpuscles with their ψ-wave. The dual aspect of light has constantly served us in this book in explaining the dual aspect of matter. Under these conditions it might seem almost certain that the theory of light should quite naturally take its place in the general framework of wave mechanics. Well! However paradoxical it may seem, it does nothing of the sort. Wave mechanics has been quite capable of establishing general relations between wave quantities and corpuscular quantities, relations with which we have dealt at length at the beginning of Chapter 8 and which are valid for photons as well as for material corpuscles, but to erect a complete theory of light on this foundation has experienced grave difficulties. Quite a few years ago, an elegant attempt was made by Heisenberg and Pauli to obtain a quantum field theory, i.e., a quantized electromagnetism which would contain in an altogether natural fashion a quantum theory of light. But this attempt, whose analytical elegance is undeniable and whose many good results will always remain, has encountered difficulties and does not seem to furnish a truly dualistic picture of light. An analogous, if not basically identical, theory was developed by Dirac and later by Fermi and others: it highlights the existence of photons better and is a very interesting theory on this account, but it does not seem to us that it is any better in realizing the desired dualistic picture.

In the face of these difficulties, some physicists have even come to doubt the existence of a real symmetry between light and matter concerning the duality of their nature. On this point we are of an exactly opposite opinion: the symmetry between matter and light, which served as the basis of the development of wave mechanics, is so satisfying to the mind and to us seems so much to be the profound reason for the success of these new theories, that, in our opinion, we must not abandon it at any price. So, we have bent our efforts these last years to bringing ourselves nearer to a truly dualistic conception of light. We shall say only a few words about this attempt now, for it is still nothing more than a venture.

One fact is undeniable: the dualistic theory of light, although having served as the model for the dualistic theory of matter, has fallen behind the newer theory. What is behind this strange fact? One of the causes is certainly the form in which wave mechanics first took in its rapid extension. This form, as we have seen, is not relativistic: it can as a consequence be applied only to those corpuscles whose velocity is small compared to that of light and can not be appropriate for photons. In addition, it contains no element of symmetry which could be used to define a polarization. Another cause, which prevents modeling the theory of the photon on that of the electron, is that the photon possesses properties which clearly distinguish it from the electron. First of all, as we know, photons, in large aggregates, obey the Bose-Einstein statistics, and not the Fermi-Dirac statistics as electrons do. And then, in the photoelectric effect, the photon disappears, is annihilated, and there exists no analogous property for material corpuscles.

From these general remarks, we concluded that in order to set up a theory of the photon it was necessary above all to use a relativistic form of wave mechanics having elements of symmetry like polarization and, secondly, to introduce *something more* in order to differentiate the photon from other corpuscles. The first part of this program is immediately realized by having recourse to Dirac's theory of the magnetic electron that we previously discussed. We know as a matter of fact that Dirac's theory is relativistic and that it has elements of symmetry which present a marked relationship with those of the polarization of light. Nevertheless, it is not enough to suppose that the photon is a corpuscle of negligible mass obeying the equations of Dirac's theory, for the model of the photon thus obtained would have, as you might say, only half the symmetry of the actual photon; in addition, it would obey, it would seem, the Fermi statistics, as the electron does and would not be capable of being annihilated in the photoelectric effect. Something more is very much needed. And this something more we have tried to introduce by supposing that the photon is made up not by one Dirac corpuscle, but by two. It can then be ascertained that these two corpuscles or demi-photons must be complementary[1] to each other in the same sense that the positive electron is complementary to the negative electron in Dirac's theory of holes (see Chapter 11,

[1] We emphasize that the word "complementary" is not taken here in the sense of Bohr.

section 5). Such a couple of complementary corpuscles can annihilate themselves on contact with matter by giving up all their energy, and this accounts completely for the characteristics of the photoelectric effect. In addition, the photon being thus made up of two corpuscles with a spin of $h/4\pi$ should obey the Bose-Einstein statistics, as the exactness of Planck's law of black body radiation demands. Finally, this model of the photon permits us to define an electromagnetic field connected with the probability of annihilation of the photon, a field which obeys the Maxwell equations and possesses all the characters of the electromagnetic light wave. Although it would still be premature to make a definitive pronouncement on the value of this attempt, it is indisputable that it leads to interesting results and that it strongly focuses attention on the symmetry properties of the complementary corpuscles whose existence, suggested by Dirac's theory, has been verified by the discovery of the positive electron.[1]

2. *Physics of the Nucleus*

Our knowledge of the nucleus of the atom has developed with a prodigious rapidity during the past few years. An entirely new science, nuclear physics of an incomparable richness, is in the process of being built up. It must therefore seem a little strange that we should be so late in approaching such an important subject. But our intention is not to outline nuclear physics and we cite two reasons which advise us not to do so. The first of these reasons is the extent of the recent discoveries made in this field: in order to give even a partially complete idea of them would require us to write a second volume or to lengthen this one beyond acceptable limits. A second reason is that our knowledge

[1] Since this text was written, the works of Dirac, Fierz, Pauli and those made at the Henri Poincaré Institute notably by Gerard Petiau, M. A. Tonnelat and ourselves, have permitted us to set up a general theory of particles with spin of which the wave mechanics of the photon such as we have just sketched it, is a special case (1946).

of the nucleus today is of a rather experimental nature. Theory has still made little progress in nuclear physics, and what has been made in this direction remains provisional. It is quite likely that the new mechanics in its turn will have to undergo modifications in order to be able to explain the behavior of the numerous elements which are found united or blended in the unimaginably small region of the nucleus. Some theories, as that of Gamow, certainly give us only a quite rough schematic picture and that most remarkable attempt of Heisenberg, too, is certainly only a rough draft.[1] In reality, nuclear physics is still at that stage of a pure and simple cataloging of facts and of establishing empirical laws: its present state is somewhat similar to that of spectroscopy before the theory of Bohr. Hence, having had the intention of writing a book directed principally toward an exposition of contemporary quantum theories, we have thought that we should devote only a final section to nuclear physics, despite its importance in the scientific movement of the present day.

We therefore wish to limit ourselves to saying a few words about the wonderful growth of our knowledge in nuclear physics, totally ignoring equally important questions such as that of isotopes, for example, or that of nuclear spin.

The nucleus of an atom of atomic number N carries, as we know, a positive charge equal to N times that of a proton and is the seat of almost all the mass of the atom. For quite some time it has been inferred that atomic nuclei are formed of protons and electrons, the number of protons surpassing by N units the number of intranuclear electrons and all the mass being due practically to the protons. The idea that the nucleus is complex is somewhat imposed by the interpretation of radioactivity.

The discovery of radioactivity, anticipated by Henri Becquerel, was the work of Pierre Curie and his collaborator and wife, Ma-

[1] The theory of Heisenberg, referred to in this text, today has been completed by the theory of the meson field, but the latter is also rather insecure in its developments. (Note added in 1946.)

rie Sklodowska, Mme. Pierre Curie. Radioactive substances are heavy elements bearing the highest numbers in the Mendelejeff table (from 83 to 92). They are characterized by the fact of being spontaneously unstable; i.e., from time to time, the nucleus of one of these atoms explodes and is transformed into a lighter atom. This decomposition is accompanied, in general, by the expulsion of electrons (β-rays), ionized helium atoms (α-rays) and very penetrating radiations of high frequencies (γ-rays). The discovery of these phenomena was of enormous interest to physicists by proving to them that nuclei are really complex structures, that by disintegrating, a complicated nucleus can give rise to simpler nuclei, thus realizing the transmutation of elements dreamed of by the alchemists of the middle ages. Unfortunately, radioactivity is a phenomenon over which we can exercise no influence and which, as a consequence, we are reduced to observing without being able to modify these processes. Hence, a great step forward was realized, when twenty years after the discovery of radioactivity, the great English physicist, Lord Rutherford, succeeded in 1919 in realizing the artificial disintegration of elements. By bombarding light atoms with α-particles (themselves emitted by radioactive substances) he succeeded in breaking the nuclei of these atoms. Simpler atoms were obtained thereby and a true artificial transmutation was effected.

[1] Since 1930 the techniques for nuclear transformations by bombardment have been perfected rapidly through the use of powerful and ingenious instruments whose prototype is the cyclotron, invented by Lawrence. In the course of these investigations, an important discovery was made by the Joliot-Curies; they showed that certain bombardment processes give rise to an unstable nucleus (artificial radio-element) which subsequently

[1] The material from here to the end of the section does not appear in the original edition and is taken from *L'Energie atomic et ses Applications* by Louis de Broglie, Mercade et Cie: Paris, 1951, p. 13-23.

disintegrates spontaneously, giving rise to another element and various radiations.

In 1931-32 nuclear physics was profoundly transformed by the discovery of two new particles: the neutron and the positive electron. The work of Bothe and Becker, of the Curies, and of Chadwick showed that the bombardment of glucinium by α-rays gives rise to a corpuscle, until then unknown, the neutron, which is electrically neutral and which has almost the same mass as the proton. The neutron has since been found in a great number of nuclear reactions and in cosmic rays.

As to the positive electron or "positron," it is a corpuscle with mass equal to that of the usual electron, but whose electric charge is equal and of opposite sign. It was discovered in cosmic rays by Anderson and by Blackett and Occhialini. The positron is unstable in the presence of matter: it has, in effect, a tendency to neutralize electrons present in matter and the simultaneous annihilation of a positron and an electron gives rise to the emission of radiation: this disappearance of a pair of electrons of opposite signs is really a "dematerialization of matter." The inverse phenomena also exists: in certain circumstances radiation can be "materialized" into a pair of electrons of opposite signs. These phenomena and other analogous ones are in accord with the principle of the inertia of energy, which only makes a change in form.

After the discovery of the neutron, Heisenberg proposed a conception of the structure of nuclei which has been shown in application to be most superior to the older one.

Nuclei would be formed, not of protons and electrons, but rather of protons and neutrons. The production of negative or positive electrons during natural or artificial disintegrations would then be explained not by the pre-existence of electrons in the nucleus as had been believed up until then, but by the transformation of a nuclear proton into a neutron or inversely, a transformation which would be accompanied by the creation of an electron. The proton and the neutron would then be somehow two states, one charged positively, the other neutral, of the same heavy particle, a "nucleon," the fundamental constituent of atomic nuclei. This body of new concepts now serves as the basis of a theory of the nucleus, and is of great service in predicting the nuclear phenomena of which we are now going to speak.

We can now make precise the use of what has been called "atomic" energy, but which should rather be called "nuclear" energy, since it is stored, not in the whole atom but in the central nucleus of the atom. For a long time man has known how to use the energy which can be liberated through the interactions between atoms at the moment when the latter combine to form new molecules or when they separate in the dissociation of already existing molecules. These transformations of the state of combination of atoms are often "exothermic," i.e., they liberate heat, energy, which we can then try to use to our advantage. The simplest example of this is certainly oxidation by combustion, "burning," whose discovery without doubt marked the most decisive turning point in the history of primitive man. The discovery of violently explosive bodies, such as nitroglycerine or T.N.T., have permitted us to obtain sudden liberations of great quantities of energy capable of producing the most violent effects. But here it is always a question of chemical energy arising from phenomena which take place on the periphery of the atoms and which modify only their mutual connections.

What then is the characteristic of this new form of usable energy which is called "atomic"? It is the fact that it springs not from the peripheral region of the atom, where the molecular bonds are made and unmade, but in that innermost region of the atom which is called the nucleus. For almost 40 years, as we have said, we have known that at the center of each atom there is a nucleus which characterizes its chemical individuality and where the major part of its mass resides. Around the nucleus in a region which, even though extraordinarily small, is extremely vast in relation to the nuclear dimensions, move the peripheral electrons. In the inner parts of this outer region of the atom the processes occur which give birth to the emissions of X-rays, while the outer parts of the same region are the origin of the visible emissions and the seat of the interactions which bring about the usual chemical phenomena. We have explained how physicists, after having been unable for a long time to make the internal structure of atomic nuclei precise, had finally understood that the nucleus is a complex system formed by an assemblage of two kinds of particles, the "protons" and the "neutrons" whose number continually increases as we consider heavier atoms and, hence, heavier nuclei. The exact structure of these assemblages, the nature of the forces which assure their stability is only beginning to appear and much progress is yet to be done toward understanding these intranuclear phenomena.

But what has been certain for not too many years is that the transformations of which nuclei are capable can liberate quantities of energy (which on a large scale, will finally manifest itself for us in the form of the liberation of heat) a great deal more considerable than those which the modifications of the peripheral state of atoms, in particular chemical reactions, can furnish. Let us note, this does not mean that the transmutation of a single nucleus will give us a large quantity of energy: this quantity of energy will be, on the contrary, always extremely small in relation to those which can usefully be employed by us, but it still is a great deal larger than what can be freed by a single process in a molecular transformation. Nevertheless, while chemical reactions have been known for a long time which abruptly give up a large quantity of energy capable of producing considerable effects on our scale, it has been only 6 years that considerable quantities of energy can be obtained from nuclear transformations. What is the cause of this?

In general, when a chemical reaction is set off in a ponderable mass of matter, it propagates itself *through* the mass. At the beginning only a few atoms of the mass are involved, but it progressively spreads, often with great rapidity, to an immense number of neighboring atoms present in the substance. The release of energy per atom is very minute, but the total release due to billions and billions of atoms becomes significant and sometimes even colossal. Now, although it had been known since Rutherford's famous experiment of 1919 how to bring about internal transformations of nuclei, artificial transmutations, the transformation of only a few nuclei in the mass of matter could be obtained. The release of energy was a great deal more considerable per nuclei than it is per atom in a chemical reaction; but, since the transmutation performed on a few nuclei does not propagate itself through the material mass, the total quantity of energy liberated remained infinitesimal.

The situation changed completely in 1938-39 through the discovery of the important phenomenon of the "fission" of uranium. Uranium is the heaviest chemical element (i.e., its atom has the greatest mass) which exists in a stable state in the earth's crust. Its nucleus, formed of 92 protons ($Z=92$) and a number of neutrons which varies from 140 to 146 according to which isotope is considered, is a complicated structure and somewhat unstable: this instability gives it a certain tendency to decompose spontaneously and this tendency results in a natural radioactivity. In 1938-39, the work of Hahn, Meitner, Strasmann, Frisch and Joliot-Curie led to a new important nuclear phenomenon: the "fission" or "splitting" of uranium. First it was noticed that the bombardment of uranium by neutrons gives rise to a disintegration of the uranium nucleus: at first it was thought that the incident neutron was incorporated into the uranium nucleus with the emission of electrons, which led to an announcement of the existence of "transuranium" elements with atomic numbers higher than 92, that would extend the Mendelejeff series beyond uranium, but would not exist normally in nature. Then other work (in France that of Joliot, principally) proved that the collision of neutrons against certain uranium nuclei broke these nuclei into two nuclei of almost equal mass, this phenomena of splitting being accompanied by an emission of neutrons. This kind of explosion of the uranium nucleus can, in addition, take place in several different ways and the nuclei which result therefrom in the different cases are themselves unstable and subsequently are transformed by emitting in turn positive and negative electrons.

After the discovery of the fission of uranium, it was believed for a while that we had been completely mistaken in our belief that the collision of a neutron and a uranium nucleus could give birth to a transuranium element. But a more thorough study of the problem led us to seeing that in reality the bombardment of uranium by neutrons gives rise to both splitting and the appearance of transuranium elements. In order to understand this we must bring in the notion of an isotope. Uranium, such as it is found in nature, is made up mainly of two isotopes, both corresponding to the same atomic number, $Z=92$. The more abundant isotope has an atomic weight of 238: its nucleus is made up of 92 protons and 146 neutrons. The other isotope of atomic weight 235, whose nucleus contains the same number of protons, 92, but only 143 neutrons is present only in the small proportion of 7/1000 in natural uranium. This rare isotope is quite unstable and it is this nucleus which explodes during collision with neutrons, producing the phenomenon of fission. As to the abundant U_{238}, its nucleus can absorb a neutron producing an uranium nucleus (still with atomic number 92) but containing 147 neutrons and having consequently an atomic weight of 239: this last nucleus is unstable and decomposes, producing an electron and a new nucleus of atomic number 93 and atomic weight 239 (93 protons, 146 neutrons). Thus a new element, which does not exist in nature, has been created and has been named neptunium. The neptunium nucleus thus formed in a mass of natural uranium bombarded by neutrons is still capable of absorbing an incident neutron, producing a nucleus of heavy neptunium with atomic number still equal to 93, but with atomic weight 240. This heavy neptunium is unstable; it decomposes producing an electron and a plutonium nucleus with atomic number 94 and atomic mass 240, plutonium thus being the second transuranium element. In résumé, the bombardment of natural uranium by neutrons gives rise both to fission of the rare U_{235} and the successive formation of neptunium and plutonium starting from the abundant U_{238}.

These facts have been known for a decade and since then we have succeeded in making other transuranium nuclei of atomic number greater than 94: these are americium ($Z=95$), curium ($Z=96$), berkeleyum ($Z=97$), californium ($Z=98$), athenium ($Z=99$) and, perhaps soon, centurium ($Z=100$)! All these nuclei are very unstable and destroy themselves by natural radioactivity. Perhaps they existed in nature at the origin of the world, but they have long since disappeared because of their natural radioactivity. In the middle of the 20th century, man has succeeded in recreating these vanished elements and it is curious that human intelligence has thus modified the natural course of the evolution of the world, at least on the surface of our planet.

Let us return to nuclear energy. The nuclear reactions of which we have spoken and which were realized just before the discovery of fission, still involved only a few nuclei of the atoms, and, despite their theoretical interest, they were only a laboratory trick without any significance for their practical utilization. However, during 1939, physicists realized that an awesome new possibility was opened to them. In effect, since the phenomenon of fission is accompanied by a liberation of neutrons, these neutrons should be able in turn to bring about the fission of neighboring nuclei. Fission should therefore be capable of propagating itself "chain-wise" to other U_{235} nuclei present in the uranium mass, if once the circumstances are favorable to this propagation. Now each act of fission liberates a quantity of kinetic energy, capable of being transformed into heat which is of the order of 3 ten-millionths of an erg, this energy being derived from the energy contained in the mass of the fissioning nucleus. This quantity of energy is very small, but if the fission spreads throughout a whole mass of uranium, the total release can be enormous. So the total fission of a kilogram of rare U_{235} can furnish (by reason of the immense number of nuclei in a kilogram) an enormous quantity of heat capable of raising the temperature of a million tons of water from 0° to 100°C. In principle, we can thus obtain an explosive a million times more powerul than any chemical explosive known, such as dynamite.

It remained only to realize this frightening possibility. Without entering into the details, let us say only that this effort followed two directions: (1.) to isolate, at least partially, the rare isotope U_{235} contained in minute proportions in natural uranium so as to obtain a body where a large number of nuclei can produce the phenomenon of fission; (2.) to make plutonium by the action of neutrons on U_{238} and then utilize the fact that the plutonium thus obtained is capable, as U_{235}, of producing the phenomenon of fission and can consequently serve in an atomic bomb. Bombs of U_{235} and of plutonium were made; the Hiroshima bomb seems to have been of the first type, the Nagasaki bomb of the second type. Since the end of the last war, these techniques have been perfected and, according to information, we are now at the point of inventing an atomic bomb of a new kind which will use the transmutation of light nuclei, such as an isotope of hydrogen. This is the famous hydrogen bomb.

Having started from the rather naïve speculations of the Greek philosophers, we have attained control by man of the energies hidden within the very heart of atoms. The possibility of utilizing atomic energy for our benefit opens a new era in the history of humanity. Human intelligence can be justly proud of having succeeded, through an intense and unflagging effort, in penetrating the secrets of the inner constitution of matter to the point of now being able to use the store of energy that it holds. From this point of view, the century-old toil of scientists, which has made them cognizant, in an ever more precise fashion, of the discontinuous structure of matter, is a sort of epic which has today culminated in its apotheosis.

Louis de Broglie: A Biographical Note

Creator of wave mechanics, Louis de Broglie is a world-famous scientist whose work in theoretical physics, combined with an estimable literary talent, has profoundly altered modern physics and put him among the foremost of contemporary scientists.

Born in Dieppe, France, in 1892, the scion of an illustrious noble family, he received his secondary education at a lyceum in Paris and in 1909 obtained a degree in history at the University of Paris; but, forsaking a career as an historian and paleographer because of his interest in science, he returned to the University of Paris and obtained a degree in science in 1913.

After military service during World War I, he worked in the laboratory set up by his elder brother, Maurice de Broglie, where he became acquainted with the experimental study of very high frequency radiations which had just become accessible to spectroscopic investigation and in which the problem of the choice between the corpuscular and wave interpretation of optical phenomena was brought into sharper focus. In 1924, Louis de Broglie submitted his doctor's thesis, *Investigations into the Quantum Theory* in which he tried to bridge the gap between these two contradictory theories. With any moving particle he associated a wave of definite wave length; but in the case of particles with a significant mass as are studied in classical mechanics, the ballistic properties predominate almost exclusively, whereas the wave properties become prominent with corpuscles on the atomic scale. Recoiling at first from the profoundly revolutionary implications of his theory, he tried by various hypotheses to preserve the traditional, deterministic interpretations of classical physics. But, because of the formidable mathematical difficulties he encountered, he was forced to align himself with the probabilistic and indeterministic interpretation wherein classical mechanics would be simply a special case of a more general mechanics, wave mechanics. Experimental verification of these theories was obtained four years later by American physicists working at the Bell Telephone Laboratories, who saw that atomic particles such as electrons and protons, because of their associated wave, would show phenomena of diffraction as light and X-rays do. Later these ideas were given a practical application in the development of magnetic lenses which are the basis of the electron microscope.

Winner of the Nobel Prize in Physics in 1929, Louis de Broglie received at the same time from the French Academy of Sciences the Henri Poincaré medal which was awarded for the first time that year. In 1933 he was admitted to membership in that body and in 1942, replacing Emile Picard, became one of its perpetual secretaries.

In addition, since 1926 he has devoted himself to matters of education and instruction. In 1928, after giving a series of lectures and courses at the Sorbonne in Paris and at the University of Hamburg, he received the chair of theoretical physics at the Henri Poincaré Institute where he created a center for study of contemporary physical theories. In 1943, anxious to overcome the problems which had been raised by the lack of cooperation between science and industry, he founded the branch of the Henri Poincaré Institute devoted to studies in applied mechanics. This interest in the practical applications of science is reflected in his recent books dealing with particle accelerators, wave-guides, atomic energy and cybernetics.

Louis de Broglie has published important scientific works on atomic particles and optics that range from an early work, in collaboration with his brother, on X-rays and gamma-rays, through the fundamental research papers on wave mechanics, to advanced textbooks in atomic and nuclear theory. In his lectures and popular books he has explored the philosophical aspects of the problems which have been raised by these new theories. His most recent work in this field is a history of the development of modern physics from the first Solvay Congress of Physics in 1911 to the present day.

His literary work won for him election to the French Academy in 1945. He is honorary president of the French Association of Science Writers and in 1952 received the initial prize awarded by the Kalinga Foundation for excellence in science writing.

When the High Commission for Atomic Energy was created by the French Government in 1945, Louis de Broglie was named technical adviser and at the time of the reorganization of the Commission in 1951, he was retained as member of the advisory scientific council.

Chronology of Important Events of the 20th Century Relating to the Development of Quantum and Atomic Theories

1901—Quantum hypothesis of black-body radiation. First appearance of the concept of quanta in modern physics (Planck)

1905—Special Theory of Relativity (Einstein)

—Explanation of the photoelectric effect by the light-quantum (photon) hypothesis. (Einstein)

1907—Quantum interpretation of specific heats (Einstein and Debye)

1910—Planetary model of the atom (Rutherford)

1913—Theoretical basis for the planetary model of the atom and interpretation of spectral lines (Bohr)

—Discovery of isotopes (Thomson)

1916—General Theory of Relativity (Einstein)

—Culmination of the "old" quantum theory (Sommerfeld, Wilson)

—Statement of the Correspondence Principle (Bohr)

1919—Artificial radioactivity (Rutherford)
1923—Discovery and interpretation of the Compton effect (Compton and Debye)
—Hypothesis of the wave character of material particles (de Broglie)
—Quantum theory for the dispersion of light (Kramers, Heisenberg)
1925—Quantum (matrix) mechanics (Heisenberg)
—Hypothesis of electron spin (Goudsmit and Uhlenbeck)
1927—Publication of the Uncertainty Relations (Heisenberg)
—Theory of the double solution and the pilot-wave (de Broglie)
—Precise formulation of wave mechanics (de Broglie, Schrödinger)
—Experimental proof of electron diffraction and the wave character of material particles (Davisson and Germer)
1928—Quantum theory of atomic nuclei (tunnel effect) (Gamow)
1930—Complete relativistic theory of the electron (Dirac)
1931—Discovery of the neutron (Bothe, Becker, Chadwick)
1932—Discovery of the positron (Anderson, Blackett and Occhialini)
1935—Postulate of the existence of mesons (Yukawa)
1938—Fission of uranium (Hahn, Meitner and others)
1942—First self-sustaining atomic chain reaction (Fermi and others)
1946—Meson field theory of nuclear radiation (Heisenberg)
1948—Artificial production of mesons (Gardner and Lattes)
1952—Revival of the deterministic interpretation of quantum processes (de Broglie, Bohm)

Bibliography

A. GENERAL REFERENCE WORKS TO THE CLASSICAL BACKGROUND

1. Maxwell, James Clerk, *Matter and Motion*, New York: Dover Publications (Reprint of earlier edition)

2. Maxwell, James Clerk, *A Treatise on Electricity and Magnetism*, Oxford: Clarendon Press, 1946

3. Einstein, Albert and Infeld, Leopold, *The Evolution of Physics*, New York: Simon and Schuster, 1938

4. Jeans, Sir James, *Physics and Philosophy*, Cambridge: University Press, 1946

5. Planck, Max, *The Universe in the Light of Modern Physics*, London: Allen and Unwin, 1937

B. THE QUANTUM THEORY

1. Gamow, George, *Mr. Tompkins in Wonderland*, New York: Macmillan Company, 1940
2. Hoffman, Banesh, *The Strange Story of the Quantum*, New York: Harper and Brothers, 1947
3. Bergmann, T., *Basic Theories of Physics: Heat and Quanta*, New York: Prentice-Hall, 1951
4. Persico, Enrico, *Fundamentals of Quantum Mechanics*, New York: Prentice-Hall, 1950 (Mathematical treatment of the theory)
5. Heitler, W., *The Quantum Theory of Radiation*, London: Oxford University Press, 1944 (Advanced treatise)

C. SPECIAL TOPICS

1. Loeb, L. B., *The Nature of a Gas*, New York: Wiley, 1931 (Kinetic theory of gases)
2. Rutherford, E., *The Newer Alchemy*, London: Cambridge University Press, 1937 (Atoms and Corpuscles)
3. Frank, Phillip, *Relativity and its Astronomical Implications*, Cambridge: Sky Publishing Corp., 1943 (Elementary Relativity theory)
4. Heisenberg, Werner, *The Physical Principles of the Quantum Theory*, Chicago: University of Chicago Press, 1930 (The Uncertainty Relations)
5. Herzberg, G., *Atomic Spectra and Atomic Structure*, New York: Prentice-Hall, 1937 (Detailed theory of spectra)
6. Coulson, C. A., *Valence*, New York: Oxford University Press, 1952

D. ADDITIONAL WORKS BY
LOUIS DE BROGLIE

1. *Matter and Light*, New York: Norton, 1937

2. *Continu et Discontinu en Physique Moderne*, Paris: 1941
3. *De la Mecanique Ondulatoire a la Theorie du Noyau*, Paris: 1946
4. *Physique et Microphysique*, Paris: 1947
5. *Optique Electronique et Corpusculaire*, Paris: 1947

Index

Q

R

S

T

U

V

W

X

Z